Breast Cancer

A Guide for Every Woman

MICHAEL BAUM
CHRISTOBEL SAUNDERS
SHEENA MEREDITH

Oxford New York Tokyo
OXFORD UNIVERSITY PRESS
1994

Oxford University Press, Walton Street, Oxford OX2 6DP

Oxford New York Toronto
Delhi Bombay Calcutta Madras Karachi
Kuala Lumpur Singapore Hong Kong Tokyo
Nairobi Dar es Salaam Cape Town
Melbourne Auckland Madrid

and associated companies in
Berlin Ibadan

Oxford is a trade mark of Oxford University Press

Published in the United States
by Oxford University Press Inc., New York

A catalogue record for this book is available from the British Library

Library of Congress Cataloging in Publication Data
Baum, Michael.
Breast Cancer/Michael Baum, Christobel Saunders, Sheena Meredith.
1. Breast Cancer–Popular works. I Saunders, Christobel.
II. Meredith, Sheena. III. Title.
RC280.B8B379 1994 616.99'449–dc20 94–1327
ISBN 0 19 262436 9

Typeset by The Electronic Book Factory Ltd, Fife, Scotland
Printed and bound in Great Britain by
Biddles Ltd, Guildford and King's Lynn

Preface

A remarkable amount of progress has been made in the field of breast cancer in the short space of time, five years, since the publication of my book, *Breast Cancer: The Facts*. It is alarming that many recommendations from that book are now obsolete. In addition my personal attitudes have changed, leading to significant changes in the text. This amply demonstrates the need, as in all areas of medicine, for patients and doctors alike to avoid dogmatism and ideology.

The other major change for this book has been the recruitment of my co-authors. Christobel Saunders is a young surgeon working in my department who has chosen to devote a substantial component of her professional life to the care of women with breast disease. She is one of the new generation of female surgeons, who are increasing in numbers, in this field as in others. I regard this as an encouraging trend in a specialty so long dominated by men. It is reasonable to say that no matter how wise and compassionate a male surgeon is, he can only imagine the significance of breast cancer to a woman. This is not to say that female surgeons should specialize in female diseases, nor that an individual patient would be better off with a female surgeon, merely that the influence of women on the thinking of the profession as a whole can lead to innovations in management and in understanding which otherwise might have been missed.

Miss Saunders' particular research interest involves assessment of the quality of life in patients being treated for early and advanced breast cancer. She has made a major contribution to these discussions within the book, as well as to the new section concerning the relationship of the contraceptive Pill and hormone replacement therapy to breast cancer risk and treatment. The inclusion of this section reflects the widespread anxiety among women about the likely effects of these drugs. While the evidence is not yet complete, what is available should help to dispel the confusion which often surrounds these issues after the publication of numerous conflicting studies.

My other co-author, Sheena Meredith, is a writer on health and psychological matters for both doctors and the lay press. She is the former health editor of a leading women's monthly magazine and for several years responded to readers' queries in its medical 'problem page'. She is well aware of the concerns of women and the prominence which both breast disease and fear of breast disease can play in their lives.

Another major revision concerns the controversial topics of breast self-examination and breast awareness, which have been the subject of much media interest recently. The section on mammographic screening has also been expanded, and incorporates new evidence emerging that a national programme of breast screening, such as that now available on the National Health Service for women over 50, can help to save lives.

There is also a chapter on breast reconstruction, which is both increasingly desired and increasingly offered to patients who have a mastectomy and, indeed, to women who have undergone this operation in the past. There are still too few facilities for this option to be available to all women who might benefit from it, although many of us are now convinced that it should be offered to all women for whom mastectomy is recommended.

Finally, our ultimate aim would be not to need a new edition of this book. Efforts are being expended in many directions to try to prevent breast cancer. These include research into the fundamental causes of the disease, attempts to detect cancer and to persuade women to report to doctors at an early stage while the disease is potentially curable, and, of course, improvements in treatment. Meanwhile however, about 25 000 women develop breast cancer each year in the UK, and many times more will have breast symptoms which they fear are due to cancer. We hope that this book will provide up-to-date information to help women to make decisions about treatment and to allay unnecessary fears and anxiety.

1994 M.B.

Contents

1

Breasts and breast cancer

Introduction

Breast cancer is one of the most important diseases of women, not only because it is both common and serious, but because, unlike many other serious conditions, it is a major concern of women even when they do not have the disease.

Most women nowadays are aware of the possibility of breast cancer. There has been a torrent of articles in newspapers and magazines, and discussions on television and radio about breasts, breast symptoms, and breast cancer.

Yet many women remain unaware of the facts about breast cancer—how common it is, what symptoms it causes, what treatment can be offered, and how successful this is likely to be. There are still many myths prevailing, and these often serve to frighten women even more. While they are right to be concerned, the facts are less harsh than many imagine. For instance, most lumps in the breast are not cancerous; treatment for breast cancer does not always involve removal of the breast (mastectomy), and for many women with early breast cancer modern treatment offers a real chance of cure. It is one of the aims of this book that women who are anxious that they might get breast cancer know what the risks are, and what really happens if the disease does develop.

Often it is only when a woman develops breast symptoms that she realizes she is unsure of their significance. Another purpose of this book is to explain the symptoms of both benign (harmless) breast disease and breast cancer, to stress that any woman with any kind of breast symptoms should always seek medical advice as soon as possible, and to describe how her doctor will try to diagnose the problem.

Other women will be reading this because they have been told that they do have breast cancer. Receiving the diagnosis is usually a shock, even for women who were expecting the worst, and often women only realize when they get home just how many questions remain unasked. On later visits they may feel that doctors in a busy clinic do not have the time to explain all the medical facts in detail, and questions often occur in between consultations. We hope that this book will also answer the questions of women with breast cancer, and provide them and their relatives with a source of accurate and up-to-date information which they can absorb in their own time.

Breast cancer and other breast diseases

Breast cancer is one of the most common cancers in women in the Western world. Because cancer of any kind tends to frighten people, and because of the emotional significance of the breast and the fear that treatment may involve its removal, breast cancer carries a particular terror. All too often this leads women to delay seeking help when they develop breast symptoms, and creates immense anxiety and distress both when awaiting the diagnosis and while undergoing treatment.

Many more women are haunted by the fear of breast cancer. In a sense this is realistic. Each year in Britain about 25 000 new cases are diagnosed and about 15 000 women die of breast cancer. Breast cancer is the leading female cancer and accounts for almost one in five of all cancer deaths among women. It ranks as the commonest cause of death overall among women aged 35 to 55. Each year one or two women in every thousand will be newly diagnosed with breast cancer. One in 12 Western women can expect to develop the disease in her lifetime, and one in 20 will die from breast cancer.

However, for every lump in the breast which is found to be cancerous, four others will prove to be benign and harmless. For every ten women with a cancerous lump, five or six can be treated initially without removal of the breast, and three or four will die of some cause other than the cancer. With early diagnosis and treatment about 85 per cent of women survive for at least five years.

So although breast cancer undoubtedly ranks as one of the most important diseases of women, the outlook is not as bleak as many women fear. In addition, in the past decade advances in treatment of breast cancer have included vastly improved understanding of its psychological consequences. No longer do women submit to operations not knowing whether or not they will awake without a breast. The options for surgical and other therapies are carefully discussed before treatment is decided and a range of counselling services and self-help groups can help women to come to terms with the diagnosis and to decide on treatment.

Surgical techniques have developed which avoid mutilating surgery wherever possible. The design and appearance of

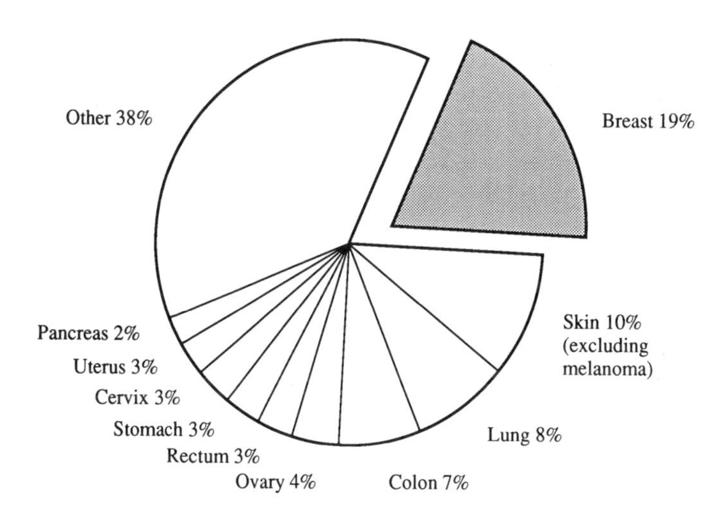

Cancer incidence among women in the UK.

prostheses have improved considerably and, if necessary, reconstructive surgery is increasingly able to offer a cosmetic result which pleases most women.

Breasts and their function

A brief description of the basic structure and function of the breast is needed to explain what happens in breast diseases including cancer. A woman's breasts are largely composed of fatty tissue, which cushions and supports the milk-producing glands. Each breast is divided into 12 to 20 lobes, each of which contains hundreds of tiny lobules where milk is produced when a woman breast-feeds. From each lobule lead tiny tubes which join up to form the milk ducts (lactiferous ducts) opening

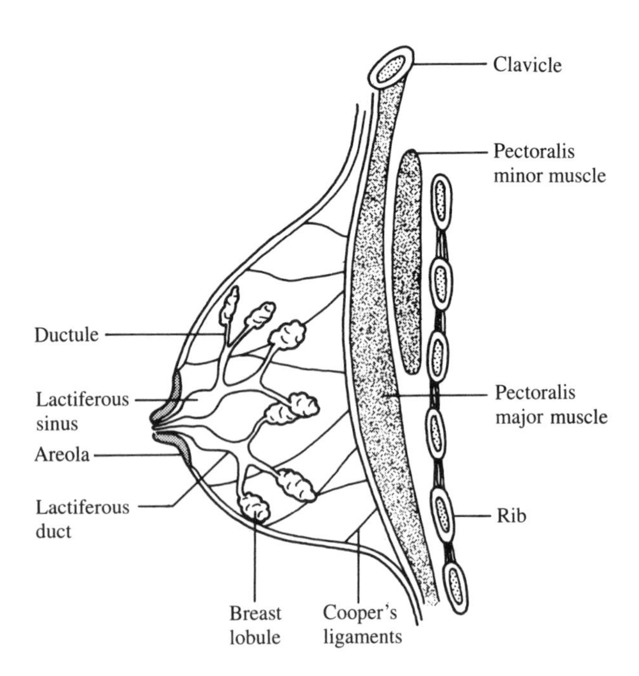

Structure of the breast.

through the nipple. Just behind the nipple each duct widens out to create a tiny reservoir called a lactiferous sinus.

Each breast is enclosed within two layers of stronger, fibrous tissue. A thin layer lies over each breast underneath the skin, and a thicker layer lies underneath the breast, between it and the muscles of the chest wall. Ligaments called Cooper's ligaments join these two layers of fibrous tissue together in between the breast lobes, and help to support the breast against the chest. These are the ligaments which may become stretched with age, obesity, or prolonged breast-feeding, so that the breasts lose their original firmness and become pendulous.

Muscles

Underlying the breast and its ligaments are the pectoral muscles of the chest wall, which lie in front of the rib cage. Pectoralis major is a large triangular muscle running between the collar bone (clavicle), the breast plate (sternum), and the sixth and seventh ribs at the front to the top of the bone (humerus) in the upper arm. Pectoralis major helps to form the contour of the

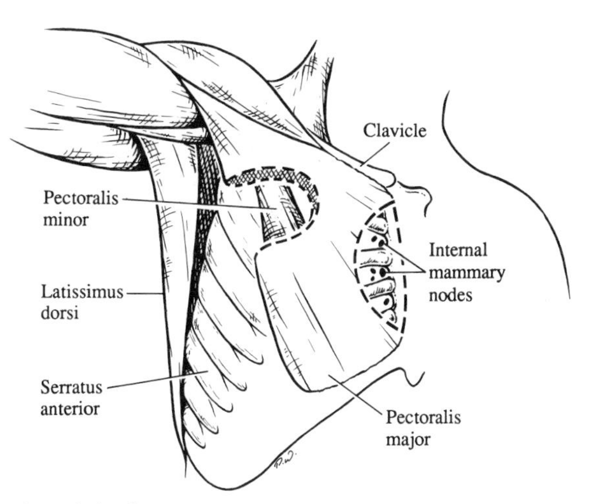

The muscles of the breast.

chest wall and is one of the muscles which acts when pulling the arm inwards towards the chest. The other pectoral muscle, pectoralis minor, is much smaller, lies underneath pectoralis major and runs between the third, fourth, and fifth ribs and the front of the shoulder blade (scapula). It helps to stabilize the shoulder girdle but is not a very important muscle.

Blood and lymphatic supply

The tissues of the breast are fed with blood, carrying oxygen and other nutrients. This arrives either from branches of a major artery in the armpit (the axillary artery), supplying the outer half of the breast, or from arteries which run between the ribs from a larger artery passing down the chest from the neck, the internal mammary artery, which supplies the inner half of the breast. Running along with the arteries are corresponding veins which carry the blood back to the heart.

The blood supply of the breast is important because cancer cells may spread through these blood vessels to elsewhere in the body. Another important means of spread is through the lymphatic system. This is a chain of vessels rather like blood vessels which drain off the fluids in which the cells, the building blocks of any tissue, are bathed. The lymph vessels drain into lymph nodes, structures about the size and shape of a kidney bean, which act as purification filters to remove bacteria and other harmful substances from the lymph fluid before it in turn drains back into the blood system.

The lymphatic system is very important in the body's defence against infection and may also play a role in controlling the spread of cancer cells or even in destroying them. However, breast cancer cells can spread through the lymph system, and many operations for breast cancer therefore involve removing lymph nodes to try to halt this spread.

The lymph flow through each breast starts around the nipple and flows outwards along with the blood vessels to lymph nodes (also called lymph glands) in the armpit or between the ribs. Thereafter it flows through various other groups of lymph nodes in the neck, around the collar-bone and breastplate.

Surgeons dealing with breast cancer try to assess whether these lymph node groups are involved or affected by the cancer as this gives important clues to the severity of the disease and the best type of treatment. The nodes in the armpit drain not only the breast but also the arm. Occasionally they are damaged either by breast cancer itself or by surgery for it, which can disrupt the drainage of lymph fluid from the arm and cause it to swell (lymphoedema).

Quadrants of the breast

Doctors describe each breast as divided into four quarters or quadrants, described as inner or outer and upper or lower, when looked at from the front. This is illustrated in the figure below. It can be seen that the breast tissue of the outer upper quadrant extends up into the armpit, or axilla. This is called the 'axillary tail' of the breast and is important because cancer can develop there as well as in the main rounded part of the breast.

Function of the breasts

Men do have some breast tissue, which can also develop cancer, although this is very much rarer than it is in women. However, the main biological function of the breasts is for milk production (lactation) to feed a newborn baby. Breasts are also an important part of a woman's body image, and have great personal and

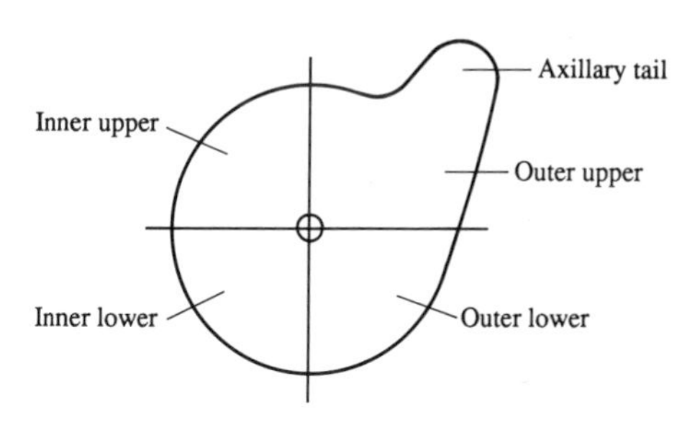

The quadrants of the breast.

social significance as symbols of motherhood, femininity, and, particularly in the Western world, sexual attraction.

Hormonal control of breast tissue

The breast tissue in women is highly sensitive to certain hormones, chemical messengers produced elsewhere within her body and circulating in the blood, which help to control the growth and development of cells, the basic building blocks of all body tissues. These hormones initiate, maintain, and control the normal physiological changes which occur during the menstrual cycle, in pregnancy and breast-feeding, and with age. As a result, the breast tissue and glandular structures are constantly changing.

Hormonal control of these processes is largely determined by the secretion of two types of hormone, oestrogens and progestogens, from the ovary. These in turn are controlled by other hormones produced in two specialized glandular structures in the brain, the pituitary and hypothalamus. Other hormones produced in the adrenal glands, which sit on top of the kidneys, the pineal gland in the brain, and possibly the thyroid gland in the neck, also play minor roles. These hormones interact in a complicated 'chain of command', with substances produced at various sites affecting the production and release of each other through complex feedback loops.

At birth and during childhood there is no difference in the size and shape of the breasts between males and females. At puberty the female hormones begin to act to initiate monthly shedding of an egg (ovulation) from the ovaries and, unless fertilization of the egg occurs, the subsequent breakdown of the lining of the womb to produce menstrual bleeding. They also cause the milk ducts and their surrounding breast tissue to enlarge. The breasts are then primed and ready for the further changes of breast-feeding should pregnancy occur. After puberty the breasts will change in size and consistency with the monthly menstrual cycle, and many women notice that their breasts tend to become more swollen, tender, and lumpy in the week or so before each period.

Breast tissue contains three types of cell which are sensitive to hormonal influences—the cells lining the milk ducts (epithelial cells), the glandular cells within the breast lobules, and specialized contractile cells found wrapped around the lobules which help to squeeze the milk out into the ducts. Each cell has on its surface specialized receptors into which hormone molecules can fit like a lock and key, so beginning a chain of reactions within the cell which prompt it to carry out specialized functions like producing milk.

If an egg is fertilized and implants in the uterus (womb), the pattern of hormone release changes. Hormones from the pituitary in the brain increase production and these, together with hormones produced by the developing placenta, influence the growth of the milk-producing glandular structures within the breast as well as all the other changes of pregnancy. The breast lobules and ducts grow rapidly in the early months after conception, but are inhibited from producing milk by placental hormones. Shortly before the baby is born the breasts prepare to produce milk, and after birth placental inhibition is removed and the baby's suckling stimulates pituitary hormones which

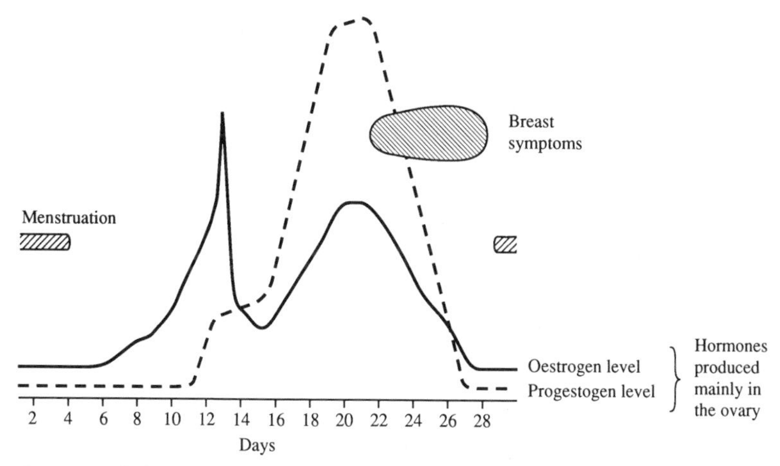

Hormonal changes with the menstrual cycle.

promote milk production and emptying of the lobules into the milk ducts.

The suckling of a baby continues to stimulate lactation hormones for as long as necessary before the baby is weaned. After weaning, or if drugs are used to stop lactation artificially, these hormones subside, milk production stops, and the breast tissue returns to its former state.

After the menopause when the ovaries stop producing oestrogens, most of the glandular milk-producing structures in the breast shrink and are replaced by fatty tissue.

What is cancer?

Cancer is not a single entity. There are many different types of cancer, and why they arise is still incompletely understood. However, the characteristic feature of all cancers is that certain cells start to grow out of control. Normal cells divide to repair and reproduce themselves only as necessary for the function of their own tissue or organ. Cancer cells keep multiplying. This is thought to be due to a change in the activity of the genes controlling these cells.

In solid tissues such as the breast or the lung, multiplication of cancer cells usually first produces a swelling, or tumour. The word tumour can be confusing because a tumour is not necessarily cancerous. If cells overgrow only in the local area where the tumour starts, the tumour is benign and usually harmless, unless the swelling presses on important surrounding structures. Benign tumours do not spread to other parts of the body. In contrast, cancerous or malignant tumours may, if not found and treated, spread beyond the initial swelling to invade and destroy the tissue around them. In addition, they may spread by the blood or lymph channels into the rest of the body, which can lead to secondary cancers elsewhere. The breast may develop both benign and malignant tumours. It is only the malignant tumours which are cancerous and dangerous.

Is breast cancer increasing?

There is a suspicion that breast cancer is increasing in Western countries, although it is only in recent years that accurate statistics have been available. Breast cancer does become more common with age and women are now living longer so are more likely to enter the age group where it is a greater risk. However, statistics comparing breast cancer rates within a particular age group today with those in the past seem to show that the figures are actually rising.

Moreover, despite improvements in treatment, the death rate from breast cancer seems also to be rising, perhaps suggesting that the disease is becoming more aggressive. Certainly among older women the disease is both more common and causes more deaths than it did 20 years ago.

Why this should be remains a mystery. However, it becomes even more important for all women, and especially those at high risk, to know which breast symptoms they should visit their doctor for, to have a full examination and, if necessary, treatment.

A note on terminology

A number of terms used by clinicians and medical researchers have a very precise definition which needs to be made plain in order to understand certain facts about breast cancer.

For instance, in ordinary conversation someone wanting to describe how common a disease is might speak in simple proportions or percentages, such as eight in ten or 80 per cent (80 in 100) of children have chickenpox. This does not tell us whether all these children have chickenpox today, or have had it in the past, or will have it before they leave school.

Doctors therefore define the occurrence of disease by two specific terms. The *incidence* of a disease is the number of new cases which arise in a specified period of time, for instance the number of new cases of breast cancer diagnosed in one year.

The *prevalence* is the number of cases found in a particular population at a given time, for instance the total number of women with breast cancer in the UK today. Each term can be made more specific by including a measure of its *rate* within a defined population, for instance the incidence per thousand women of childbearing age.

As well as how common a disease is, there are two words used to describe how detrimental it is. *Morbidity* refers to the sickness or handicap caused by a disease, complication, or treatment, for example a sore shoulder after an operation for breast cancer. *Mortality* refers to deaths attributable to these causes, and the *mortality rate* means the death rate over a specified period (usually annual) among a defined population, usually of a specific age range, for example, the number of women dying of breast cancer each year per 100 000 women over the age of 50.

Various means of studying the incidence, prevalence, morbidity, and mortality of diseases, and for comparing treatments and outcome, are used. *Epidemiological studies* which look for differences between populations are discussed in Chapter 11. *Clinical research* generally looks at individuals to compare those with a particular condition or having a particular treatment (*cases*) with unaffected or untreated people (*controls*).

For instance, clinical research can compare women with breast cancer with unaffected women in *case-control studies*, trying to find differences between them in terms of, say, their medical and reproductive histories. Or large populations of currently healthy women may be assessed and followed-up in a *prospective study* to try to find differences between women who develop the disease and those who do not. *Cohort studies* are a particular form of prospective study which follow groups of otherwise comparable women distinguished by one feature in which the researchers are interested, for example Pill-users and non-users. An *intervention study* tries to alter characteristics thought to be associated with disease and see if this makes any difference, for example comparing breast cancer rates in women on a particular diet with control women.

Finally, the most important means of assessing the effects of treatment is through an intervention study in the form of a *randomized, controlled trial,* which is often *double-blind*. This is the best means doctors have found to be absolutely sure that a treatment truly has the benefits proposed, and that any adverse effects are discovered. It is the basis on which most drugs given a licence to be prescribed in the UK are tested, and is designed to eliminate any bias which might allow errors to creep into the results.

Randomization means that patients are assigned to the study groups completely at random, for example by numbering patients in the order in which they are seen in a clinic and putting even numbers in one group and odd numbers in another. Other means of selecting patients would risk the groups being uneven, for example there might be a tendency to select patients with more severe disease for an experimental treatment, and the results could then suggest less of an effect than if a truly representative selection of patients were used.

Controlled means that patients receiving the drug or treatment under assessment (the active group or treatment group) are compared with control patients not so treated.

Double-blind means that neither the researcher nor the patient knows which treatment is being given. This may be achieved, for example, by having identical pills, one of which contains an active drug and the other an inert or placebo substance, and having an independent third person put the tablets in bottles labelled only with a code number. It avoids either the doctor or the patient prejudicing the results because of their expectations of improvements or side-effects.

A double-blind technique is of no use for studying treatments whose results are obvious, such as comparing operations for breast cancer which conserve the breast with those which do not, but it may be important in assessing certain drug treatments. An example is the trial looking at the value of tamoxifen in prevention of breast cancer in women at high risk (see Chapter 11).

Very clear ethical limits are placed on what can be studied in these types of trial, and no drug is given to patients until it has been thoroughly tested in the laboratory, on animals and on human volunteers. Patients must give *informed consent* to such a trial, meaning that they understand that they are taking part in a study, they know the effects of the proposed treatment and that it will be given (or not) at random. It is through patients generously agreeing to take part in such studies that genuine advances in treatment can come about.

2

Breast symptoms

The most common breast symptoms which lead a woman to consult her doctor are a lump, nipple discharge or bleeding, and pain in the breast. Each of these could be a symptom of breast cancer, but is much more likely to be due to some benign and harmless condition. Any woman who notices any unusual symptom or change in her breasts should see her doctor as soon as possible, although most will not have cancer. All the diseases mentioned by their medical names in this chapter will be described in more detail in Chapter 3.

'Normal' breast symptoms

It is worth mentioning several changes which can occur in a woman's breast which are perfectly normal but which may cause concern. Many women, especially in their thirties and forties, experience variation in size and 'lumpiness' of their breasts, sometimes with aching or pain, due to the hormonal changes of their monthly menstrual cycle. The breasts can also become tender to touch. Typically this occurs in the last few days before a period, and the symptoms are relieved as soon as the period begins. Such problems can start in adolescence but are much more common in the mature woman.

Swelling of the breasts occurs in pregnancy, and painful engorgement is common just after a woman has given birth, while her milk is 'coming in', but its cause is then readily apparent, as is the soreness of cracked nipples which can occur in the early days of breast-feeding. In the first few days after giving birth the breasts secrete a thin, watery fluid called colostrum before they begin producing milk proper.

Women's breasts vary enormously in size and shape according to their inheritance and there is little other than surgery which can be done to alter their contour. Many women have one breast which is slightly bigger than the other, and for a few women the discrepancy between the two is quite obvious, and may cause the nipples to be at different levels. As long as this is long-standing it is not a cause for concern except for cosmetic reasons.

Most women with perfectly normal breasts have small swellings around the nipple, called Montgomery's tubercles. These are modified sweat glands and are quite harmless, but they often enlarge during pregnancy and breast-feeding. They can also become more prominent during sex, and may be tender for a while afterwards, which is also quite normal.

Other permanent, though not strictly normal, differences in the appearance of the breasts are absent breasts and extra breasts. In rare cases a baby may be born with no breast tissue at all. On the other hand some women have a developmental abnormality in which extra breast tissue forms along the so-called 'milk line', which runs from the armpit to the groin, and is a developmental remnant of the multiple pairs of mammary glands of other mammals. This extra tissue may take the form of a complete extra breast which behaves just like a normal breast and will even produce milk after giving birth. More commonly it appears as an extra nipple, which may seem no more than a mole or blemish. In theory, these supernumary breasts may also develop cancer. Curiously, they are more common in men.

Of course trauma to the breast cannot be considered 'normal' either, but many women are very anxious about blows to the breast, whereas they are little different to blows elsewhere on the fleshy parts of the body and hardly ever do any lasting damage. A hefty knock may hurt and make the breast extremely tender, and extreme cases cause a nasty bruised swelling called a haematoma, but it will not cause cancer. The reason blows to the breast have been linked with cancer in many people's minds is probably that a woman who finds a lump and has had a recent knock to the breast is likely to associate the two, whereas such blows are easily forgotten if nothing untoward is noticed soon afterwards. Very occasionally a blow to the breast may lead to a condition called fat necrosis, in which a hard lump forms when the fat cells are disrupted.

In many women even when not pregnant or breast-feeding, a thin, milky-type discharge occurs at a very low rate. A slight accumulation of this can create a cream-coloured, curdy debris at the end of the nipple where the milk ducts disgorge. This is quite innocuous, as is the off-white discharge which can occur due to hormonal changes around the menopause.

Any soreness or redness of the nipple should immediately be reported to a doctor, but one possible cause worth considering in women who play sports is the much-publicized 'jogger's nipple'. This results from rubbing on clothing and features a tender nipple from which the top layer of skin has been excoriated. In extreme cases this can bleed.

Lumps in the breast

The vast majority of breast cancers are first discovered when a woman, her partner, or doctor finds a lump in the breast. A lump is the first symptom in 80 per cent to 90 per cent of breast cancers. However, the vast majority of lumps in the breast are not cancerous—80 per cent are found to be due to harmless conditions, most often a benign tumour called a fibroadenoma, or a cyst (a fluid-filled swelling which feels very round, smooth, and hard). Although most breast cancers occur after the age of 40, a lump in the breast should never be ignored and no woman, whatever her age, should feel tempted to wait and see if it goes away.

Although a doctor can gain useful hints from clinical examination it may not be possible even for an experienced breast cancer specialist to tell for sure simply by feeling a solitary lump whether it is cancer or not. Typically, cancerous lumps are single, hard, painless to touch, and may have irregular edges. However, benign conditions, such as a fibroadenoma, may also feel like a single, hard lump.

The typical cancerous lump as it presents today is much smaller than tumours presenting in the past, at about 2 cm in diameter by the time it is large enough to feel, and it is most noticeable when feeling the breast with the flat of the hand. The upper, outer quadrant of the breast is the most common site, with 60 per cent of cancers arising in this section, but any part of the breast tissue can be affected.

Benign lumps include fibroadenomas, which can arise anywhere in the breast and are often multiple. They usually occur in young women in their late teens and twenties. Typically they move away from the hand when touched, hence doctors often refer to them as 'breast mice'. They usually grow over two or three years to up to 2 cm in diameter and then gradually shrink.

Fibroadenomas are part of a group of conditions now termed 'ANDI' or 'anomalies of normal development and involution'. These used to be called by a range of often confusing terms including fibrocystic disease, chronic mastitis, and fibroadenosis. ANDI includes a range of breast disorders from the normal lumpiness usually felt before a period through to well-defined lumps and cysts.

Other possible causes of breast lumps are rare (less than 5 per cent of all lumps), and include fat necrosis, other cysts, and various swellings of the chest wall which are discussed in Chapter 3. Statistically, a solitary lump in the breast of a woman over 55 is likely to be cancer. A woman under the age of 30 is likely to have a fibroadenoma, and in between some form of cyst is the most likely cause.

Bleeding or discharge from the nipple

Bleeding from the nipple is a rare but understandably alarming symptom. It is not a common feature of breast cancer, but since this is one possible cause it should always prompt urgent medical attention, particularly if the bleeding is from one side only. Less than 3 per cent of women with breast cancer notice bleeding as a first symptom.

In a small proportion of women with bleeding, a lump in the breast can be found (although this may only be discovered on careful examination by a doctor), and this increases the likelihood of a cancer being present. If a lump is not found, bleeding or a blood-stained discharge is more likely to be due to a duct papilloma, a tiny, wart-like growth within the milk ducts.

Discharge from the nipple which is not bloodstained is rarely due to breast cancer. Usually it is associated with duct ectasia, a benign condition of menopausal women in which the milk ducts swell.

During early pregnancy as the breasts are preparing themselves for their future milk-producing role there is sometimes a clear discharge, and very rarely this may be bloodstained. Provided there is no associated lump this requires no treatment and should cease after the birth of the baby.

A milky-coloured discharge is common after breast-feeding and also with a type of cyst called a galactcoele which contains milk. Occasionally medicines which stimulate hormone production can produce a milky discharge, in men as well as women. The most common are the contraceptive Pill, some drugs which lower blood pressure, and some antidepressants. The discharge should resolve when the drug is stopped although this may take a few weeks.

A yellowish or brown-green discharge can occur in chronic inflammation of the breast ducts, known as periductal mastitis; a complication of duct ectasia. In addition there may be a tender red area beside the nipple, which may produce pus, and scarring underneath the nipple may pull it inwards.

Pain in the breast

Pain in the breast is a common symptom which is rarely due to cancer, although traditionally breast cancer was believed to cause a 'pricking' pain. Apart from the 'normal' cyclical breast pain mentioned above, which may be sufficient to warrant treatment with tablets, the most common cause of pain in the breast is an abscess, which can cause intense pain until relieved.

Sometimes pain is not due to breast disease at all but to conditions of the chest wall, which may feel as if the pain is coming from the breast. Examples would be inflammation of the costal cartilages which join the ribs to the breastplate, or a pulled muscle. In older women pain coming from the heart in angina may occasionally be felt as if it was in the breast.

Other symptoms

Any breast tumour, whether benign or cancerous, which grows to a considerable size may cause the affected breast to become noticeably larger. Typically a woman may notice an imbalance in the size of the breasts, or that one has become more pendulous, or that her nipples are at different levels. In advanced breast cancer however, the affected breast may shrink as the tumour replaces normal breast tissue with scarring. Sometimes a woman may notice a change in the shape of the breast before she detects the lump which is causing it.

Sometimes when a lump in the breast is discovered, a woman will have hard lumps in her armpit as well. Occasionally a lump in the armpit may be the first one to be noticed. This is classically a symptom of breast cancer, and suggests that the cancer has already spread to the lymph nodes in the armpit. However, lumps under the arm may also occur in benign breast disease and with infections. Even with breast cancer, swollen lymph nodes do not always indicate that the cancer has spread to them, as sometimes they may enlarge as part of an immune reaction by the body, presumably in a natural attempt to resist the cancer.

Because tumour cells are multiplying, they need an increased blood supply, and some women may notice that the veins running across their breasts become more prominent.

Some women notice indrawing or dimpling of the skin of the breast. This is highly suggestive of breast cancer, since a malignant tumour may attach to surrounding tissues within the breast. Skin puckering with breast cancer is due to infiltration and shrinkage of Cooper's ligaments (see Chapter 1) and accumulation of fluid in the tissues (oedema). In extreme cases this can lead to a pitted, orange-peel appearance of the skin, which doctors call 'peau d'orange'.

A cancerous lump may be clearly attached to the overlying skin (tethering) or to the muscles underneath the breast tissue. Skin tethering rarely occurs with benign disease, but is occasionally seen in fat necrosis and periductal mastitis.

Similarly, sometimes with breast cancer the nipple becomes inverted, pulled to one side or elevated because of the tug from within the breast although, of course, some women have inverted nipples as a perfectly harmless condition, and the nipple may be retracted with duct ectasia.

Symptoms of advanced breast cancer

If a woman does not seek medical attention and the cancer is allowed to become advanced, eventually the lump, and sometimes lymph nodes under the armpit which are involved, can grow to a large size and involve the skin. An ulcer may then develop which looks unsightly and may become infected and nasty-smelling. This is called ulceration and fungation. In some advanced cases the lymph circulation may be blocked and the arm may swell, a condition known as lymphoedema.

In advanced disease which has spread around the body there may be general symptoms, or symptoms of secondary cancers, in addition to symptoms of disease in the breast itself. The symptoms of advanced disease are discussed in Chapter 9.

Symptoms which may be due to breast cancer

- lump in the breast (most common)
- change in size or shape of the breast

- pain
- indrawing or inversion of the nipple
- rash around the nipple
- bleeding or discharge from the nipple
- dimpling of the skin of the breast
- lump in the armpit
- prominent veins on the breast
- swelling of the arm
- ulceration of the skin
- symptoms of secondary tumours elsewhere

3

Benign and cancerous changes

As explained in Chapter 1, a tumour arises when certain cells lose their normal control of growth and division. The cells therefore continue to divide and multiply (replicate) even when this is no longer necessary for repair and replacement, effectively becoming immortal. In solid tissues, like breast, bone, or lung, this creates a swelling, which is the literal meaning of the word 'tumour'.

There are many different kinds of solid tumour arising from different sorts of tissue, and they behave in very different ways according to the type of tumour (benign or malignant), the kind of cells, and the site involved.

Benign tumours do not spread beyond their site of origin and are therefore usually harmless, unless the swelling causes deformity or presses on important surrounding structures. They are hardly ever life-threatening and can almost always be treated and completely cured.

Malignant tumours are cancerous, which means that they are able to overcome the normal controls which stop cells in one area invading other tissues. If left unchecked they will invade and destroy surrounding tissue and may eventually spread through the rest of the body to cause secondary cancers elsewhere.

Doctors sometimes use euphemisms when talking to each other in front of patients, to avoid needless alarm, for example when discussing the need to rule out cancer. The most common euphemisms for cancer are neoplasm or neoplastic (meaning new growth) and mitotic lesion (meaning a disease process arising from rapidly dividing cells).

The process of spread of cancer cells occurs via the blood or lymphatic channels and is called dissemination or metastasis. The original tumour is known as the primary, and tumours which arise after spread of cancerous cells elsewhere are called secondaries or metastases. The rate at which cancerous tumours invade or spread varies enormously, and this determines how malignant and therefore how dangerous the tumour is. Aggressive cancers can spread very rapidly and kill the patient within weeks of diagnosis, whereas many slow growing cancers are only discovered by chance after someone has died of something else.

The different types of tumour are usually named after the medical term for their tissue of origin. For example, benign tumours of glandular tissue are called adenomas, and those of fibrous tissue are called fibromas. Thus in the breast the most common benign tumour arises from glandular tissue and also contains fibrous elements, so is called a fibroadenoma.

The most common cancers are those which arise from epithelial cells (which make up the surface layers of skin and mucous membranes and the lining of internal body surfaces) or glandular tissue, and these are called carcinomas. The most common breast cancer arises from the cells lining the milk ducts and is known as a ductal carcinoma.

Benign breast disorders

Benign disease is by far the most common cause of breast problems in women. Up to 30 per cent of all women need some form of treatment for benign breast disease at some time in their lives. Yet benign disease receives little attention in the media and has largely been ignored by researchers, because it is so overshadowed by breast cancer.

When any woman goes to her doctor with breast symptoms, the first and most important consideration for both of them is likely to be to exclude breast cancer. Once this has been done any benign condition can be diagnosed and, if necessary, treated. It is very important that any woman with a benign

Classification of benign breast disorders

- Anomalies of normal development and involution (ANDI)
  - painful, lumpy breasts
  - fibroadenomas
  - cysts
- Duct ectasia
- Epithelial hyperplasia
- Other benign tumours
  - lipomas
  - duct papillomas
- Inflammations and infections
  - trauma
  - mumps
  - breast-feeding
- Galactocoele
- Fat necrosis
- Congenital disorders
  - inverted nipple
  - supernumary breasts/nipples
- Non-breast disorders
  - skin conditions eg sebaceous cysts
  - conditions of the chest wall

---

breast disorder who develops a new lump, or any other breast symptom, should not assume that this is a further manifestation of the same disease. Since both benign disease and breast cancer are common, they may coexist. As with any lump in the breast a woman should see her doctor to be sure of the diagnosis.

In the past a wide variety of terms have been used to describe the form of benign breast disorder now known as ANDI and these have often generated confusion among both doctors and patients. The classification has now been considerably simplified in a manner which enables easier understanding of the underlying processes of the disorders. The term 'benign breast disorder' may be used for specific breast diseases and for other non-cancerous conditions such as developmental abnormalities of the breast (see Chapter 2) and inflammations.

ANDI

ANDI are the most common forms of benign breast disorders. Many of the symptoms of ANDI are similar to those which women experience as part of their monthly cycle, and which are so common that they may be regarded as normal. Because breast tissue changes so much in the normal course of events (see figure below), with the menstrual cycle, with pregnancy and breast-feeding, and with age, there can be a fine borderline between normal physiological changes and established abnormalities warranting the diagnosis of 'disease'. Grouping these conditions together under the term ANDI enables us to see them as a spectrum from extreme cyclical changes at one end to clearly defined abnormalities, such as fibroadenoma, at the other.

ANDI are probably due to excessive or abnormal responses within the duct and glandular tissue of the breast to normal circulating hormones. The term includes fibroadenoma, the condition formerly referred to as fibrocystic disease (or chronic mastitis, or fibroadenosis), other breast cysts, and cyclical nodularity of the breasts.

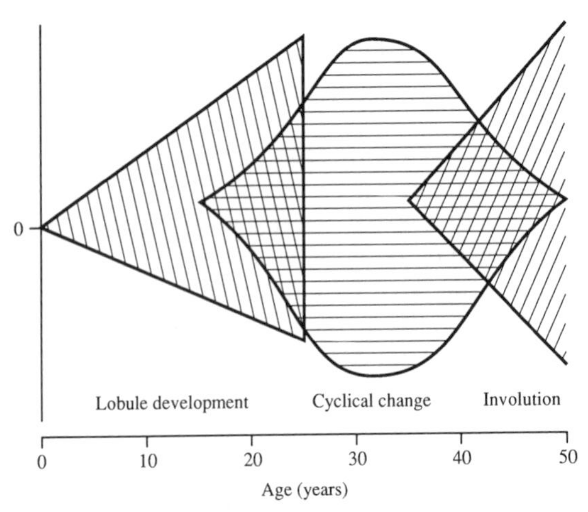

Normal breast changes throughout life.

The disorders grouped under the ANDI umbrella can largely be divided by the age at which a woman is affected. Shortly before puberty some girls develop an obvious lump underneath one or both nipples. This is simply a variant of normal and no treatment is required except to reassure the girl.

Fibroadenomas

Fibroadenomas are benign tumours of the fibrous and glandular tissue within a breast lobule. They arise in the fully developed breast, typically between the ages of 15 and 25 although occasionally in much older women, and are extremely common. They produce a well-defined, painless lump which feels like a marble and slips away from the hand when pressed. They grow very slowly, usually to about 2 cm in diameter, and then generally shrink.

Fibroadenomas are generally harmless, and in a woman under 25 will not usually require treatment unless they become very large or tests suggest some reason for suspicion. If a woman is over 25 the lump will usually be removed surgically to be sure that it is benign. This procedure is usually done under general anaesthetic and leaves only a small hairline scar.

Breast pain and nodularity

Painful, lumpy breasts are most common in women in their thirties and forties, but may occur at any time during reproductive life. In about 40 per cent of women pain is worst just before menstruation. In a mild form this can be considered a normal response to fluctuating hormone levels, and often no treatment is required. If the symptoms are bothersome, wearing a well-fitting bra, or taking mild painkillers, often helps considerably. Both the contraceptive pill and hormone replacement therapy have been linked with this kind of pain, and changing the type of tablet may cure the problem. Another link is with a high intake of caffeine (in tea, coffee, or cola), and cutting down often helps.

In about 5 per cent of women these measures are not enough to control the symptoms and drug treatment is needed. Oil of evening primrose taken for at least three months is effective in many women and has virtually no side-effects. If this does not work then drugs such as danazol, bromocriptine, or tamoxifen may be given. Finally, some breast pain can be found to have a 'trigger point' which may be injected with local anaesthetic.

Cysts

Cysts most often occur in the last decade of a woman's reproductive life, as the breast tissue is shrinking before the menopause. Often several occur at once, and they may be in both breasts. Cysts containing large amounts of fluid can feel solid and may mimic a cancer, so they are usually aspirated, that is, the fluid is drained out through a syringe, as a diagnostic procedure which also abolishes the cyst. However, in about 30 per cent of cases cysts will recur.

Duct ectasia

This is a condition of older women, typically occurring around the time of the menopause. Its cause is unknown but it has been linked with smoking. The milk ducts become inflamed and fill with a sticky yellow, sometimes bloodstained secretion which discharges from the nipple. If the condition becomes chronic fluid may leak into the surrounding breast tissue, spreading the inflammation, a condition called periductal mastitis.

Often both breasts are affected. The area around the nipples may become red and itchy, sometimes with pain and local tenderness. Sometimes an abscess forms and discharges onto the skin around the nipple, and this can become a persistent feature known as a mammillary duct fistula. In extreme cases the skin may be drawn in and nipple retraction can occur.

Since periductal mastitis can mimic malignancy a specialist opinion should always be sought and a mammogram will often be performed to exclude cancer. The symptoms often resolve by themselves, but may require antibiotics or, in advanced cases, a minor operation.

Epithelial hyperplasia

This means an overgrowth of the cells lining the milk ducts. It is common among premenopausal women, and can cause lumpiness or a nipple discharge, but is usually found by chance during a breast biopsy for some other reason. Provided cancer is excluded by biopsy, no further treatment is needed.

Lipomas

Lipomas are benign tumours of fat cells. They can occur anywhere in the body, and are not uncommon in the breast since it contains a high proportion of fatty tissue. They can form large fatty lumps under the skin, so are often removed for cosmetic reasons (a minor operation with little scarring), but are otherwise painless and unimportant.

Duct papillomas

A duct papilloma (also called intraduct papilloma) is a tiny, benign, wart-like growth of one of the milk ducts just under the nipple. It is so small that usually no lump can be felt and typically it only comes to light because pressure on the duct produces a yellow or bloodstained discharge from the nipple. Occasionally a pea-like lump will be felt under the nipple.

Duct papillomas can occur at any age but become more common in older women. They are harmless, but should always be removed because very rarely they may undergo further changes to become malignant. Removal can be carried out through a small operation under general anaesthetic which usually does not affect the appearance of the breast or the nipple. Very occasionally on examination in the laboratory an intraduct papilloma will be found to contain cancerous cells and then more extensive surgery will probably be advised.

Inflammations of the breast

Acute inflammation of the breast (acute mastitis) arises in reaction to trauma, infection, or irritation of the breast tissue, which becomes hot, red, swollen, and painful. This is unlikely to be mistaken for cancer, and there is usually an obvious precipitating cause. Mild versions may arise in the newborn and at puberty, probably as a result of hormonal influences, and occasionally a traumatic inflammation can arise where straps rub and irritate the breast area. Rarely, mumps can lead to acute mastitis, and more often it may stem from infection of one of Montgomery's tubercles. Infections may also follow duct ectasia, as discussed above.

However, by far the most common cause is an infection during breast-feeding, which unless treated promptly is likely to produce an abscess. This classically affects mothers who are just starting to breast-feed, in the first month after their first birth, when sore and cracked nipples may allow the entry of bacteria. It can be excruciatingly painful, particularly if combined with swelling or engorgement of the breasts.

Frequent feeding combined with expressing milk either by hand or with a breast pump usually gives substantial relief. If caught before an abscess has formed the condition may be cured with antibiotics, but if an abscess is present the pus will need to be drained by a minor operation.

Nowadays most doctors encourage women to carry on breast-feeding on the affected side provided the drainage incision can be kept away from the baby's mouth (and in any case feeding can continue normally from the other breast).

Galactocele

This is a type of cyst which occasionally develops during pregnancy, and causes a smooth, tender lump which fluctuates in size. It is due to blockage of one of the ducts which then becomes swollen with milk. It usually resolves when breast-feeding stops.

Fat necrosis

This is a condition which is claimed to occur after a blow to the breast, leading to a hard, irregular lump and sometimes to swollen lymph glands under the arm. This can mimic cancer, which should always be excluded unless the lump shrinks away rapidly.

Skin conditions

Since the breast is covered in skin, it can develop the same benign skin conditions which occur anywhere on the body, such as sebaceous cysts, due to blockage of the grease-producing glands of the skin. These are completely harmless but may be removed (excised in medical parlance) if the woman dislikes their feel or appearance.

Conditions of the chest wall

For completeness mention should also be made of those conditions which may mimic breast disease but actually arise in the chest wall. These include blood vessel swellings, tuberculosis or tumours affecting the ribs, and inflammation of the costal cartilages which join the ribs to the breastplate, a condition called Tietze's syndrome, which is simply treated with rest and mild analgesics.

Breast cancer

Up to now we have generally talked about breast cancer as if it was a single condition. In fact, several types of cancer can arise in the breast. The most common by far, as mentioned at the start of this chapter, is the duct carcinoma, followed by carcinoma of the breast lobules, and then rare forms of cancer. In addition, for both duct and lobular carcinomas there is a localized form of pre-invasive disease, referred to as *in situ* carcinoma, which may or may not progress if untreated to 'full-blown' cancer. To distinguish them, the latter are therefore referred to as invasive duct cancer and lobular invasive carcinomas, invasive in this context meaning that the tumour has invaded outside the duct or lobule in which it arose to the surrounding breast tissue, not necessarily to tissues outside the breast. These are illustrated in the figure on p. 33.

Invasive duct carcinomas

These arise from the cells lining the milk ducts and comprise over 90 per cent of all breast cancers. The first symptom is generally a hard, somewhat ill-defined lump within the breast. Gradually as the tumour spreads along the ligaments between the breast lobes it pulls on the overlying skin to create a characteristic dimple. If the tumour spreads along the ducts these will pull on and eventually invert the nipple. Lymph nodes under the armpit may be involved, and as the tumour spreads it may involve the skin, the underlying muscles or even the structures of the chest wall. Extreme skin pitting (peau d'orange) is a grave sign, as is skin ulceration. The smaller and less advanced the cancer is at the time of diagnosis the better the outlook for the patient.

Lobular invasive carcinomas

Much less common than duct carcinomas are those which arise in the lobules of the breast. Lobular carcinomas account for about 8 per cent of breast cancers and behave in a very similar way to duct carcinomas, except that often there are several sites within the same or the opposite breast affected with either invasive or pre-invasive disease at the same time.

Pre-invasive disease

For carcinomas of both duct and lobular origin, a pre-invasive form of the disease may occur. These are known as intraduct cancer or duct carcinomas *in situ* (DCIS), and lobular carcinomas *in situ* (LCIS) (*in situ* being the Latin term meaning confined to its original situation).

Duct and lobular carcinomas *in situ* are confined entirely within the duct or the lobule in which they originated, and show no evidence of invasion of the surrounding breast tissue. A lump is rarely detectable and unless there is some other symptom such as nipple discharge the majority come to light through mammographic screening of the breast.

In situ carcinomas pose a problem for treatment because it is not known what proportion of cases progress to invasive cancer, nor how long they may take to do so. However, samples removed for laboratory analysis suggest that as many as 40 per cent have the potential (shown by initial stages of invasion through the lining of the duct or lobule) to become malignant, and the risk of invasive breast cancer developing in a woman who has had DCIS treated by removing the affected area alone is believed to be about 11-fold increased.

On the other hand, post-mortem studies suggest that some 5 per cent to 6 per cent of apparently normal breasts contain

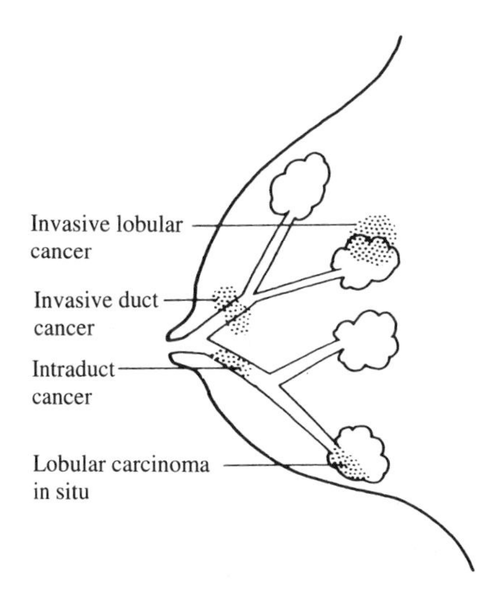

Types of breast cancer.

in situ carcinomas, and this proportion rises to as much as 40 per cent in breasts removed at mastectomy because of a cancer detected elsewhere in the breast tissue. Some studies of women with breast cancer have suggested that the opposite, unaffected breast may contain DCIS in up to 60 per cent of cases. Clearly not all of these progress to invasive cancer, but there is at present no means of telling which ones are likely to do so.

Paget's disease of the nipple

Occasionally breast cancer comes to light after the development of what appears to be eczema of the nipple in a middle-aged or elderly woman. A red rash around the nipple progresses to scabbing and ulceration of the nipple and areola, and may be confused with benign conditions such as eczema. In this case however the symptoms are due to an underlying carcinoma affecting the skin around the nipple. Although this is rare, any woman with such symptoms should always have a biopsy to exclude cancer.

Rare forms of breast cancer

Much less common is a type of breast cancer arising from the connective tissue cells (found in cartilage and ligaments) of the breast. Connective tissue tumours are known generally as sarcomas, and in the breast a fibrosarcoma, with both fibrous and connective tissue elements, is occasionally seen.

Finally, the breast may be the site of secondary cancers which have spread from primary tumours elsewhere in the body. This is uncommon but most likely to be from tumours of the chest wall which have simply extended into the breast, or from a type of skin cancer, the melanoma or malignant mole.

Spread of breast cancer

As already mentioned, breast cancers may spread locally, direct to surrounding tissues, or distantly through the blood or the lymphatic systems. The lethal propensity of cancer results from its ability to shed single cells or clumps of cells at distant sites in the body, where they may develop into secondary cancers.

The most common sites for early spread of breast cancer are the lymph nodes draining the breast. If laboratory analysis of a sample of lymph nodes from the underarm or from the chain of nodes running behind the breastplate shows invasion by cancer cells, this is regarded as a grave sign. Cancer cells may spread from these nodes through the lymphatic system to other node groups, so later in the course of the disease the lymph nodes around the collarbone may be affected, and later still those within the chest or abdominal cavity, the groin, or even the opposite armpit.

The lymphatic system plays an important role in the body's immune system, that is, its defence against infection, toxins, and, perhaps, cancer. The lymph nodes may be important in the body's attempts to counteract breast cancer. It is now thought that if the lymph nodes do not contain cancer cells this may indicate a high natural resistance on the part of the patient. On the other hand, involvement of the local lymph nodes may suggest that natural resistance is exhausted, and hence that there is a strong chance that the disease has spread elsewhere in the body. These factors may have an important bearing on treatment.

The lymphatic vessels and the veins draining the breast communicate in places, so it is probable that either route of 'escape' of breast cancer cells could enable them to spread anywhere in the body. However, breast cancer spread via the blood stream is probably the most important in determining the final outcome of the disease.

In recent years evidence has accumulated suggesting that breast cancer cells enter the bloodstream relatively early in the course of tumour development. This is why treatments are now commonly used which aim to eradicate cancer cells from the body as a whole, rather than just dealing with the primary tumour.

Blood spread probably occurs as a result of direct invasion by tumour cells of the veins draining the breast. Any viable cancer cells which spread around the blood stream may seed out to form secondary deposits, most commonly in bone, liver, lungs, brain, or ovaries. Such secondary tumours show that the disease is widespread and may directly contribute to a fatal outcome through, for example, liver or respiratory failure.

In advanced disease the membranes surrounding the lungs (pleura) and gut (peritoneum) may become involved. This enables fluid leakage in these areas which can progress to large collections of fluid (pleural effusion and ascites, respectively), which are uncomfortable and distressing signs of terminal disease.

4
Diagnosis

Clinical diagnosis

Most women with symptoms of breast disease first see their family doctor, who will make an initial assessment. If it is necessary, a woman may then be referred to a specialist in the local hospital for out-patient tests. This is most likely to be a surgeon with a special interest in breast care.

The initial assessment will probably involve questions about the symptoms themselves, about the woman's breasts, her general health and family history, followed by an examination of the breasts and underarm area.

Examination by a specialist

A full examination of the breasts involves first looking at the patient. The woman is generally asked to sit, first with her hands by her sides and then with her arms raised above her head. This is so that the doctor can look for asymmetry between the breasts, for nipple retraction or indrawing, and for skin dimpling.

The woman may then be asked to lie back for the doctor to examine her breasts, carefully feeling each quadrant of both breasts with the flat of the hand. The essential point of this examination is to decide whether there is an obvious, separate lump, or whether the breasts are simply generally lumpy. The underarm region on both sides will then be checked for any lumps caused by swollen lymph nodes. If a breast lump is found the next stage is a test aspiration (see following text) to see whether it is cystic or solid.

If a woman's symptoms are of nodularity of the breasts or pain or both, but no obvious lump is found on examination, further management will depend on her age. For women under 40 such a picture is so unlikely to be associated with cancer that it is reasonable to adopt a wait-and-see approach, and she may be asked to return in six weeks. If a lump is then found test aspiration can be performed. If (as is more likely), there is still no lump, she can be reassured that there is no evidence of cancer. If the symptoms have by then resolved, no further action is necessary; if she still has painful or lumpy breasts then treatment may be advised, either with mild painkillers or oil of evening primrose (see Chapter 3).

For a woman over 40 in the same situation, a mammography (breast X-ray) may be advised to ensure that no hidden cancer is present. If this shows no evidence of malignancy she can be reassured and offered mild treatment for the symptoms if necessary. If there is any suspicion of a cancerous pattern on the mammogram, a needle localization biopsy will probably be advised (see Chapter 13).

If the woman has complained of a discharge from the nipple, the doctor may ask her to demonstrate this by massaging the area under the areola to express some fluid, which can then be tested to see if it contains blood, as this is not always obvious to the naked eye. The doctor will pay special attention to whether the discharge comes from a single duct or many, as this determines what happens next (see figure on p. 39).

Depending on the woman's history and what has been found so far, the examination may proceed further, perhaps assessing the area around the collarbone for lumps, listening to the chest, and feeling the abdomen to check for signs of spread of disease elsewhere.

If there is an obvious lump in the breast the next step is to biopsy it. This can be done using either a fine needle and syringe (fine needle aspiration cytology) or a slightly larger cutting needle.

Fine needle aspiration

An ordinary hypodermic needle is inserted into the lump and the syringe attached. If the lump is a cyst it will be filled with fluid and this may be drained completely. If the cyst does not recur, and as long as there is no blood in the cyst fluid, no further treatment is needed and the procedure both confirms the diagnosis and effectively cures the cyst.

However, in the few cysts which do recur or contain blood there is a small chance (about 1 in 2000) that a cancer is present in the cyst itself. In these cases an open biopsy (see following text) may be recommended.

If the lump is solid a few cells can be withdrawn into the needle and these sent for microscopic examination using special staining techniques—known as cytology.

The great advantage of fine needle aspiration cytology is that

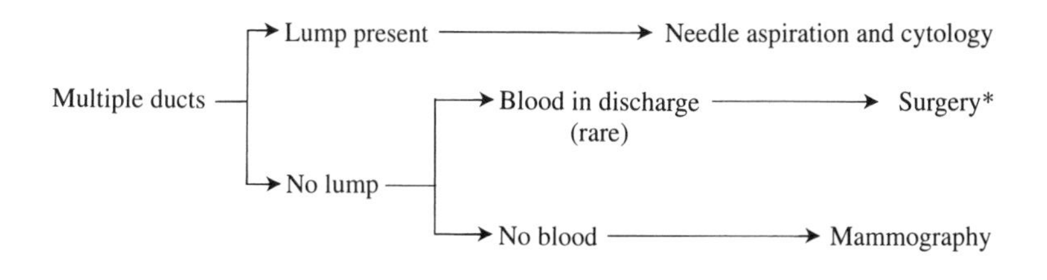

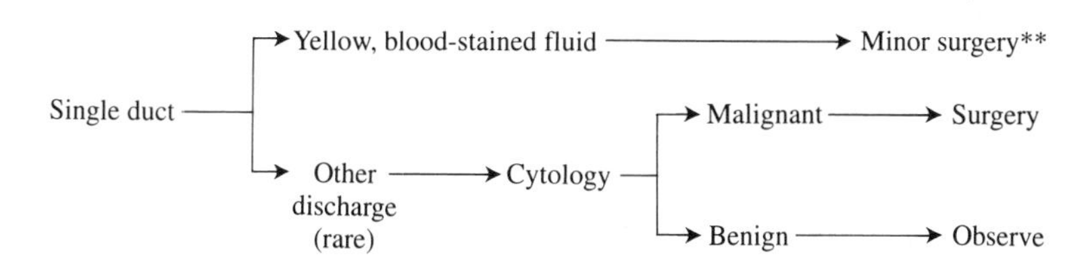

* Removal of a wedge of tissue behind the nipple containing the terminal milk ducts (Hadfield's operation), used to treat severe cases of duct ectasia.

**Removal of a single milk duct (microdochotomy) thought to contain a duct papilloma.

The diagnosis of nipple discharge.

in expert hands the examination of the sample can take place within a few minutes, while the patient is still at the clinic. If cancer cells are found, a definitive diagnosis at the first visit can be of enormous benefit in sparing the woman the traumatic wait until the next out-patient visit, and enables the surgeon to begin discussions with the woman about choice of treatments.

Cutting needle biopsy

This is sometimes used if the cytology results are not conclusive or if the specialist requires more than just a few cells to make a diagnosis. Under local anaesthetic to numb the skin, a special needle capable of slicing and withdrawing a small specimen of tissue is passed into the lump. This sample may be analysed in the laboratory, a technique known as histology.

The disadvantage of both of these techniques is that a negative result does not always prove that there is no cancer, as the area sampled may not have been the one containing the cancer cells. If there is any suspicion of cancer either on clinical examination of the lump or on its X-ray appearance, an open biopsy will be performed.

Mammography

Mammography, or breast X-ray, may be used as a diagnostic as well as a screening test. A cancer may reveal itself by a specific pattern in the tissues surrounding a cancer as well as by the presence of an obvious lump. Cancer is obviously the main thing a doctor will be looking for, but other diseases can be diagnosed on mammography too. However, although it has an accuracy of over 95 per cent for diagnosing clinically detectable breast lumps, this is not sufficient in dealing with possible breast cancer, so for any obvious lump or other suspicious changes found, test aspiration or biopsy is generally recommended.

If both tests prove negative the lump alone can be removed at a minor operation and the woman reassured. Sometimes in a woman under 30 the specialist can be so sure that a lump is benign that the woman can be offered reassurance alone. If either or both tests show malignancy, surgery for breast cancer can be commenced.

As discussed above, mammography may be recommended for women over 40 who have no obvious lump but whose breasts have a coarse, nodular texture which could mask a relatively small malignant tumour. It is also useful in some cases of nipple discharge. In addition, since some forms of cancer feature multiple cancers or *in situ* carcinomas, mammography may be useful to detect other tumours in the same or the opposite breast.

If no abnormalities have been detected by a doctor's examination of the breasts and a mammogram shows no areas of suspicion, a woman can be reassured that her symptoms are almost certainly not due to cancer.

There are two other indications for mammography which do not fall under the screening umbrella. One is to follow-up a woman after treatment for breast cancer, since she is at risk of recurrent disease and has a higher than normal chance of developing a new breast cancer. The other is where an abnormality within the breast tissue needs to be accurately localized before surgery. This technique and the general principles of mammography are discussed in Chapter 13.

Ultrasound

Ultrasound involves the use of high frequency sound waves passed through the body. Structures of different density bounce these sound waves back more or less strongly, and the resulting signals can be converted to images on a screen which experienced operators can interpret.

Ultrasound scanning can help to distinguish a solid lump from a cyst, which can be useful when the lump cannot be accurately localized and the woman's breasts are too dense for mammography to be used, which is often the case in young women. Unfortunately it is not good at detecting small abnormalities, so is not very useful as a screening procedure for early breast cancer. It is sometimes used in advanced disease to look for secondary tumours elsewhere, for example in the liver.

Open biopsy

Unless she has had cancer demonstrated unequivocally by needle biopsy (and thus can proceed straight to definitive surgical treatment) any woman over 30 with an obvious breast lump should have it removed. The same applies if she has suspicious nodularity or mammographic findings. In all cases the aim is not to miss a case of cancer at a stage where it might be curable.

Removal of a lump is called open biopsy (because the sample is taken at operation, rather than through the skin) or excision biopsy (because the whole lump is cut out). It is carried out in hospital under a general anaesthetic.

In the past, a procedure called 'frozen section' was commonly used to assess the character of the lump. This meant that slices of tissue were cut and fast frozen to enable immediate laboratory assessment. The result was relayed back to the operating theatre while the woman was still under anaesthetic, so that if the lump was cancerous surgical treatment (then generally mastectomy) could be carried out under the same anaesthetic.

Not surprisingly, not knowing whether or not she was going to wake up minus her breast made this a particularly traumatic procedure for the woman, and meant that many women with benign lumps had to face the trauma of consenting to mastectomy even when this was not carried out. Nowadays, unless the clinical or mammographic findings make the surgeon fairly certain that a cancer will be found, open biopsy is performed purely to remove the lump and send it for analysis, and if cancer is found the surgeon discusses the treatment options with the woman and her family before she submits to a second operation.

Laboratory analysis

Biopsy or operative samples sent for laboratory analysis are first, if they are large enough, examined with the naked eye (macroscopically) and then specially prepared by slicing and staining to be looked at under a microscope. The study of tissue samples like this is known as histology. For tiny samples such as those obtained by fine needle biopsy, or for fluid drained from cysts or samples of nipple discharges, special procedures for the study of cells (cytology) are undertaken. These specialized techniques can distinguish features of both benign diseases and the various forms of cancer with a high degree of precision.

An adenocarcinoma, for example, has a quite characteristic appearance. Usually a cut surface of the tumour looks grey, granular, or gritty to the naked eye, and has radiating spicules giving the crab-like appearance from which cancer earns its name. Under a microscope the cancer can be clearly seen as pockets of abnormal cells scattered through the fibrous matrix which gives the cancer its hard feel. This is known as a scirrhous pattern, from the Greek word for hard, and is by far the most common.

Two much rarer subtypes of adenocarcinoma are those in which tubular or mucus-secreting cells predominate. These are not so different to the normal cells of origin of the tumour, so are called well-differentiated tumours. They carry an excellent outlook.

Conversely, a cancer may contain cells which are quite unrecognizable as coming from breast tissue at all. These are known as undifferentiated tumours and carry a much poorer prognosis. The degree of differentiation is known as the grade of the tumour and can give a fairly reliable guide to the outlook for the patient irrespective of the clinical stage of the tumour (how advanced it is).

In addition to the cancer cells themselves the tumour may

contain an influx of cells from the body's immune system. This is generally believed to reflect the body's natural reaction to the presence of cancer cells. However, in one type of breast cancer, called a medullary carcinoma, specialized cells of the immune system called lymphocytes are in the majority. Again, to the experienced eye this will be evident as a characteristic pattern under the microscope.

A biopsy sample taken at operation will usually include some of the tissue surrounding the tumour, as well as the cancer itself. If the blood and lymph vessels in the local area of the tumour also contain cancer cells, this may predict the ability of the tumour to spread to other sites. This can be a very important prognostic indicator for an individual cancer.

Spread of the disease

Once breast cancer has been diagnosed and its type established in the laboratory, there are a variety of other important considerations which help the specialist to determine how severe the cancer is and what is the best form of management.

Node status

One of the first things an examining doctor will look for after feeling the breasts themselves is the state of the regional lymph nodes. The nodes under the armpit (axillary nodes) are usually the first to be affected, and the degree of involvement may vary. Initially the nodes may feel enlarged and hard, then gradually as they grow they, like the primary tumour, may invade surrounding structures and so become fixed to deeper structures in the armpit, or invade and eventually ulcerate the overlying skin. The lymph nodes above and below the collarbone (supra- and infraclavicular nodes) may also be involved.

Involvement of the regional lymph nodes is one of the first signs that the disease is spreading, and if cancer cells are found in biopsy samples of the axillary nodes this is generally taken as indicating a worse prognosis. Nowadays it is believed that axillary involvement is simply a chronological indicator of how long the disease has been present, rather than a specific risk factor on its own account, and that the biological features of the tumour cells themselves may have a greater independent influence on outcome. However, there is no doubt that involvement of the axillary nodes is associated with high probability of widespread disease at the time of diagnosis and therefore with a worse outcome. The more nodes are involved the worse the prognosis seems to be.

Because of this, whether enlarged nodes are found on clinical examination or not, and even if the nodes are not removed altogether, samples are often taken from the axillary area during surgery for breast cancer. These can then be examined in the laboratory to check for evidence of node involvement, which may guide the need for adjuvant treatment.

Biological tumour markers

It has become increasingly obvious recently that how an individual tumour will behave, and thus the outcome for any patient, depends not only on the tumour size and presence of lymph node involvement, but also on a number of characteristics of the tumour itself.

The tumour characteristic most commonly assessed is the oestrogen receptor (ER) status. Breast cancer cell growth is dependent on oestrogens, the 'female hormones', produced largely by the ovaries (but also in other tissues such as fat later in life). This fact forms the basis of many treatments for breast cancer such as drugs aimed at abolishing the secretion of oestrogen. Some breast cancers are more sensitive to oestrogen than others, and contain specific areas called oestrogen receptors on the cell surfaces to which molecules of oestrogen attach. Tumours with the receptors are more likely to respond to antihormone treatments. However, this is by no means the whole story as some tumours without oestrogen receptors respond to antihormones, and many tumours become resistant to these drugs despite having abundant receptors.

Other biological factors measurable in tumours can give clues as to how aggressive the tumour is likely to be in terms of growth rate and potential for spread. Unfortunately, so far no one test has proved a very useful predictor of outcome for the individual patient, but it is hoped that more work on these factors in the future will allow doctors to tailor treatment to an individual woman and her tumour.

Diagnosis of secondary spread

To check for spread of the disease elsewhere, even if there are no obvious secondary tumours, a series of simple tests are generally performed during the initial assessment of a woman with breast cancer. These include a chest X-ray to look for secondaries in the lungs or involvement of the membranes around them, and a blood test to look for anaemia or abnormalities in the blood cells, which may indicate bone marrow involvement, or for any abnormalities of blood chemistry which could indicate bone secondaries or involvement of the liver. If a woman has no clinical evidence of secondary tumours, and her blood tests are normal, the chances are that she has only local disease.

In more advanced cases a variety of specialized tests are used to check for secondaries, including X-rays or bone scans of other likely sites of bone spread (skull, spine, pelvis, and hip) and ultrasound to assess the state of the liver. These tests are discussed further in Chapter 9. However, it is worth noting here that they are not infallible. Tiny clumps of cancer cells (called micrometastases) may already have seeded out in other organs but cannot be detected by conventional methods. This is probably why some cases progress to widespread disease even after aggressive treatment of the primary tumour.

Staging the disease

There is no doubt that the smaller and less advanced the tumour is at the time of diagnosis, the better the survival prospects for the patient irrespective of treatment. Some assessment of the likely stage of the disease can be made on purely clinical grounds after a woman's initial examination.

Broadly speaking, in Stage 1, the disease is confined to the breast, with or without minor skin dimpling or nipple indrawing. In Stage 2 the regional lymph nodes are involved as well but are not fixed to the surrounding tissues. Stages I and 2 are potentially curable and therefore aggressive surgery and other treatments with toxic side-effects, such as chemotherapy or radiotherapy, are justifiable. Clinical examination is not completely reliable in differentiating Stages 1 and 2.

Stage 3 disease is locally advanced breast cancer, where either the tumour or the lymph nodes have infiltrated the skin or the underlying muscles or other structures and are clearly

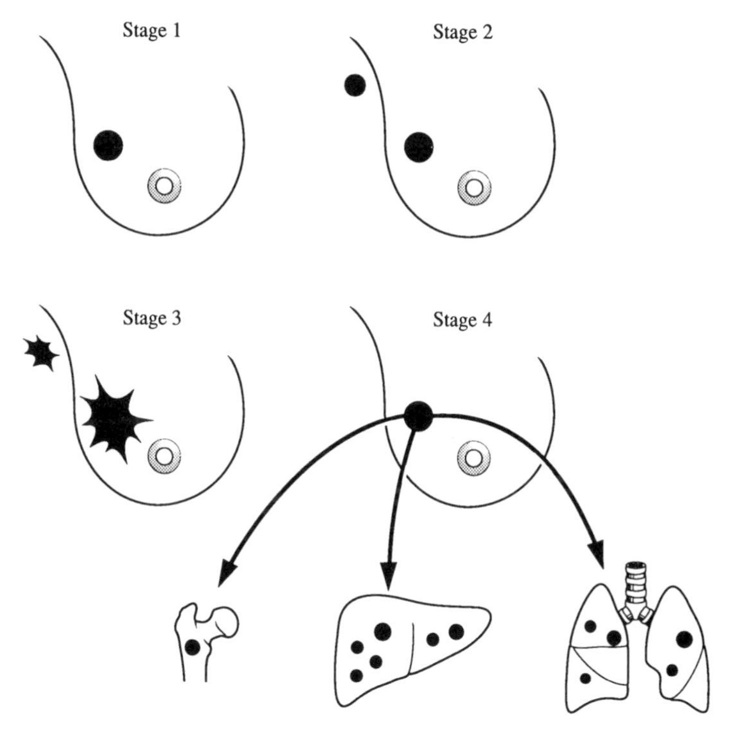

Staging of breast cancer.

fixed to them. In Stage 4 there is obvious secondary spread to elsewhere in the body. Stages 3 and 4 are classified as advanced disease, where the aim of treatment switches from cure (which is now unobtainable) to symptom relief.

Some surgeons take a more pragmatic approach to staging. Women are assessed as minimal risk if they have a tumour less than 1 cm in diameter with no involved lymph nodes, or if a small tumour is of one of the rare tubular or mucinous subtypes with a better outlook. Moderate risk involves tumours between 1 and 2 cm with less than four involved axillary nodes. Any other category is high risk. In this system a woman assessed as minimal risk has a 95 per cent chance of surviving 10 years after local surgery, and will rarely require other treatments with toxic side-effects.

Clinical staging can be further refined by what is known as the TNM system, standing for tumour, nodes, and metastases (secondaries). This staging system is recommended by the International Union Against Cancer (UICC), so is sometimes known as the UICC system. As an example, a small tumour confined to the breast with no evidence of lymph node involvement or secondary tumours elsewhere would be classified as T1 N0 M0 on this system, whereas a larger cancer with involved lymph nodes in the armpit but still without secondaries might be classified as T2 N1B M0.

Other systems which originated in the 1940s to distinguish operable tumours from more advanced cases are the Manchester staging system, in which Stages 1 and 2 represent operable and potentially 'curable' cases, Stage 3 locally advanced disease, and Stage 4 obvious spread of the disease with secondary cancers, and the Columbia clinical classification, where Stages A and B are defined as operable and potentially 'curable' and Stage C represents locally advanced disease.

Diagnosis of recurrent disease

All women who have had breast cancer should be carefully monitored for several years after treatment, to ensure that there is no recurrence of the disease. Usually a woman is asked to attend for an annual out-patient visit for at least five years after treatment, and in many cases for life. During these visits she will be asked about any symptoms (although of course if symptoms return in between visits they should be immediately reported to her GP), her breasts will be examined and other tests, such as mammography or bone scans, may be done if necessary.

5

Treatment by surgery

Until very recently the surgical treatment for breast cancer was so standard that in many women's minds mention of the disease was almost instantly equated with mastectomy, the removal of the entire breast. In the past decade surgery for breast cancer has undergone a dramatic change, and the aim is now to conserve the breast whenever possible.

'Conservative' surgery which removes only the lump or the segment of the breast containing the tumour, leaving most of the breast intact, is becoming increasingly accepted as a safe option, and in the UK is now offered to up to two-thirds of women with small tumours. Removal of just the lump is popularly known, despite objections to the hybrid derivation of the word, as 'lumpectomy', -ectomy being the suffix used in medicine to mean cutting something out at surgery (mastectomy is from the Greek *mastos* for breast and *-ektome* for cutting out). If an entire quadrant of the breast is removed the operation is known as quadrantectomy or segmentectomy. All these options may be referred to as wide local excision. Lumpectomy usually retains both a breast mound and the nipple and areola, so clearly gives the least alteration to the woman's appearance.

Conservative surgery is the most common means by which breast cancer can be treated without removal of the entire breast, but there are other options which may be more suitable in certain cases, in particular subcutaneous mastectomy, which removes the tissue of the breast while leaving the skin and nipple intact. Very rarely a tumour may be treated by radiotherapy or chemotherapy alone.

Generally lumpectomy is offered to women with early breast cancer, that is, a small, solitary tumour confined to one breast without evidence of spread except to the axillary lymph nodes on the same side and with no involvement of skin, muscle, or lymph nodes elsewhere in the body. It is also emerging as the preferred surgical treatment for *in situ* disease, although some surgeons may recommend subcutaneous or simple mastectomy.

If by the time breast cancer is diagnosed the tumour is too big or has spread too far to be treated by local removal alone, mastectomy may be needed. Women with relatively large tumours compared with the size of the breast, or those whose tumour is situated immediately behind the nipple, may need mastectomy to achieve adequate local control of the disease.

Some women choose mastectomy, perhaps because it is the only way they can feel confident that the cancer has been entirely removed, or perhaps because they do not want adjuvant (additional) treatment such as routine radiotherapy which is more often recommended after lumpectomy. So even today up to half of all women with breast cancer still have a mastectomy.

Irrespective of what type of operation they have, almost all women having surgery for breast cancer are suitable for some form of reconstructive procedure (see Chapter 6) where necessary to restore a pleasing cosmetic appearance.

Breast conservative surgery

The advantages for an individual woman of retaining her breast are obvious. In addition, many people feel that the more widespread use of lumpectomy could have benefits for all women, since if they are aware that breast cancer need not necessarily imply mastectomy, they may be encouraged to see a doctor at the earliest opportunity should they discover a breast lump.

There are no absolute guidelines as to which women are suitable for lumpectomy rather than mastectomy. Treatment choice will be determined for each woman after careful discussion with her doctor, taking into account such factors as the size of the lump in relation to the size of the breast, the site of the tumour, its stage and grade, as well as the woman's preferences on both surgery and adjuvant therapy. In most cases adjuvant treatment will be recommended in addition, so the woman needs to take into account how she will cope with this, since it is likely considerably to lengthen the period of treatment.

The operation itself is generally a minor procedure carried out under general anaesthetic and will usually involve a stay of four or five days in hospital to ensure that the scar is healing well and there are no after-effects.

Most surgeons either remove the lymph nodes under the armpit (axilla) on the same side as the tumour at the same time as lumpectomy, or advocate postoperative radiotherapy to the axilla for these patients, to try to ensure that any cancer cells which have spread to these nodes are eradicated. If the lymph nodes are removed, it is not a good idea to give radiotherapy to the axilla as well, as this can lead to swelling of the arm (lymphoedema).

Radiotherapy to the breast area itself may still be given in

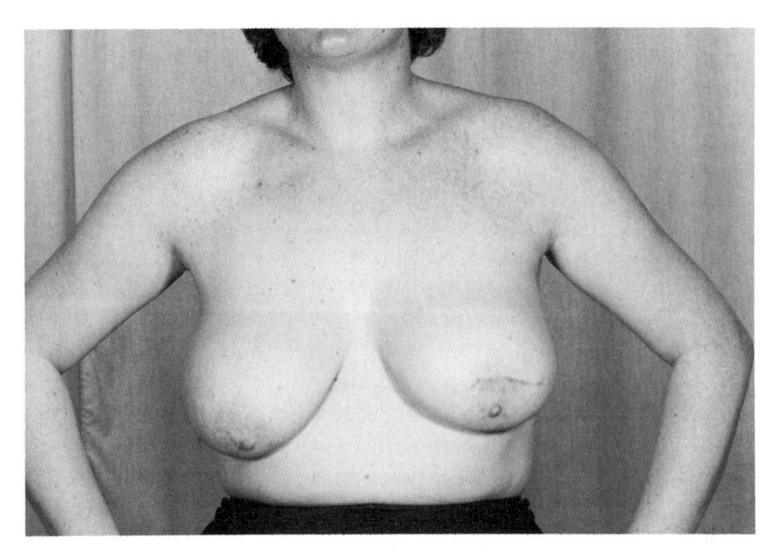

Appearance after lumpectomy.

these cases, and is commonly advocated after lumpectomy whether the axillary nodes are removed or not, to prevent local recurrence of the disease.

Often a tumour is small enough for a lumpectomy to be carried out without noticeable effect on the volume of the breast, and scarring is minimal. With larger tumours, and particularly if a whole segment of the breast is removed (quadrantectomy), there may be a significant loss of breast tissue.

In most cases the nipple is unaffected by conservative surgery, and breast-feeding may be possible provided the milk ducts have not been damaged by radiotherapy.

Mastectomy

With the exception of subcutaneous mastectomy, which leaves the skin over the breast intact while removing the breast tissue, mastectomy involves removing the entire breast including the overlying skin with the nipple and areola. There are various types of mastectomy which differ in the structures removed in addition to the breast itself.

Mastectomy is a fairly major procedure. Most patients stay in hospital for about a week after the operation, depending on the type of surgery.

The cosmetic result varies according to what structures are removed. Clearly breast-feeding will not be possible from the affected side, but if a woman does have a baby after mastectomy it is usually possible to feed from the other side alone, and the other breast tends to increase its milk supply to compensate.

Subcutaneous mastectomy

It is possible to remove most of the breast tissue under the skin but to leave the skin and nipple intact, combining this operation with immediate reconstruction with an implant. This yields a fairly good cosmetic result because the outside of the breast is still the woman's own.

However, it is suitable only for a few selected cases of

breast cancer, and is more often used in women with high risk pre-cancerous conditions such as *in situ* carcinoma.

Simple or total mastectomy

This removes all the breast tissue including the extension of the breast into the armpit (the axillary tail) and the nipple and areola and surrounding skin. An oval cut (incision) is made to remove the breast tissue and closed as a transverse line, leaving a flat chest with a horizontal scar, sometimes with a curved extension up into the armpit. Usually surgeons now also remove a few samples of the lowermost axillary lymph nodes to send for laboratory analysis.

Modified radical (Patey) mastectomy

This removes all the structures taken away in the simple mastectomy, and also the pectoralis minor muscle, the small

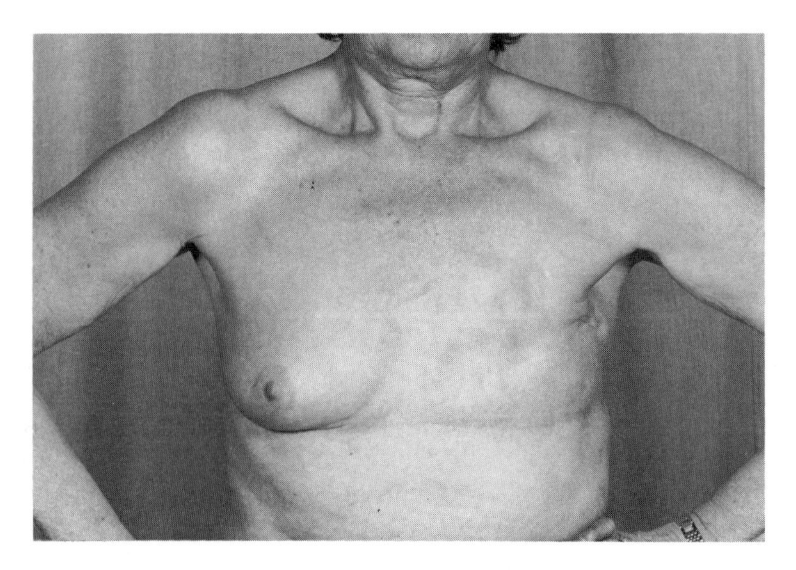

Appearance after simple mastectomy.

and largely insignificant muscle crossing the armpit from the ribs to the front of the shoulder blade. Removing pectoralis minor as well allows easy access to the armpit, so that all the lymph nodes and surrounding fat pad can be cleaned out too, but makes little difference to the bulk of tissue removed. The incision is closed in the same way, with a transverse line, and the results are indistinguishable from simple mastectomy, allowing women to wear swimming costumes and low necklines with an external prosthesis.

Classical (Halsted) radical mastectomy

This operation, devised in the late nineteenth century by a Baltimore surgeon, may be regarded as the start of modern surgical techniques for breast cancer. Until 20 years ago it was the most commonly performed operation for breast cancer. It no longer has a place in routine treatment, but is described here as many women will have had this operation in the past.

It removed the pectoralis major muscle as well as all the structures removed in the modified radical mastectomy. This meant that the shape provided by the muscle at the front of the chest wall was lost, so after the operation the woman was left not only without a breast but also with a hollow beneath the collar-bone where the ribs showed through. This created a considerable deformity visible under all but the most concealing of clothes.

Even more extensive operations were sometimes performed in the past but are now never used.

Removal of lymph nodes

As explained above, many operations for breast cancer involve removing the axillary lymph nodes as well as the tumour. Surgeons differ in their approach to this because the importance of removing these nodes is still debated. It is now believed that lymph node involvement is a sign that the disease may have spread widely within the body. Surgical removal might therefore be important for symptom relief, but will not necessarily increase the cure rate. Certainly, whether or not the nodes are removed seems to have little effect on overall survival rates, and radiotherapy to the axilla may be just as effective in controlling the local disease.

Whether or not the nodes are removed as part of the treatment, samples of them will frequently be taken for laboratory analysis to aid in determining the stage of disease. This is important both in assessing the likely outlook for the woman and in deciding whether or not chemotherapy is indicated, particularly in premenopausal women. In postmenopausal women sampling of the nodes is less important as decisions about non-surgical treatment can often be made without knowledge of the axillary node status.

Timing of operation

A curious result which emerged from a recent study is that the timing of surgery for breast cancer may affect its outcome. Among premenopausal women, those operated upon during days three to 12 of the menstrual cycle (day one being the first day of a period) had on average a 54 per cent chance of surviving for ten years, whereas this increased to 84 per cent after surgery at other times. The difference was most marked among women with involved axillary nodes, for whom surgery in the second half of the cycle more than doubled the ten-year survival rate to 78 per cent, compared with 33 per cent among such women operated on in the first half of the cycle.

The explanation is probably hormonal, as during the early part of the menstrual cycle oestrogen levels are high and progesterone (another hormone which to an extent antagonizes the effects of oestrogen) levels are low. It could be that any cancer cells released into the body during surgery are more likely to survive in a particular hormonal environment. However, other studies attempting to confirm this finding have not shown the same effect, and at present any influence of timing of surgery for breast cancer must be regarded as unproven.

Surgery for advanced breast cancer

Even where disease is so advanced that it is no longer possible to aim for a cure of breast cancer, surgery may be performed to deal with the local tumour and its effects, in particular skin ulceration and involvement of the chest wall. Surgery may also be needed to relieve symptoms of secondary tumours. These options are dealt with more fully in Chapter 9.

Side-effects of breast cancer operations

Apart from the changes to a woman's appearance, there are a number of other possible adverse consequences of breast cancer surgery.

In the immediate postoperative period, there may be complications affecting the area around the scar, including bruising, wound infection, swelling, and failure of the scar to heal properly. Usually such complications will be detected and dealt with before a woman leaves hospital. However, if a woman has chemotherapy at around the time of her operation, this can delay the onset of wound infection by several weeks.

With all forms of mastectomy there is likely to be some degree of weakness of shoulder movements afterwards, and these must be built up again by postoperative exercises.

Swelling of the arm due to accumulation of lymph fluid (lymphoedema) where the lymph nodes and vessels are removed or damaged can be extremely painful and awkward to manage. Fortunately this complication is now less common as radical procedures are performed less often and the combination of removal of the axillary nodes with axillary radiotherapy is now avoided.

Occasionally the nerve supply to two muscles in the base of the armpit is damaged during radical surgery, which can result in instability and 'winging' of the shoulder blades. Again, this is now rare as radical surgery is not often performed.

Evidence for the effectiveness of different operations

Despite all attempts to reduce mortality with different types of surgery for breast cancer, the chance of survival seems to be determined more by the stage and grade of the disease at the time of diagnosis than by what form of surgery is performed.

Attempts to obtain accurate comparisons of survival rates after different operations for breast cancer have been problematic. Firstly, very large numbers of women must be studied to reveal significant differences between procedures, and they must be followed-up for a long time (which means keeping track of them over years and perhaps decades), since breast cancer can recur many years after apparently successful surgery. Secondly, the outcome may be influenced by additional treatments such as radiotherapy or chemotherapy, which makes assessment more complicated. Thirdly, there have been ethical doubts about the justification of randomizing women to less extreme forms of surgery which, it was feared, might compromise survival.

Lastly, there has been a particular difficulty in conducting studies of mastectomy versus breast conservation. To remove the uncertainties about choice of treatment, a formal randomized trial is crucial (see Chapter 1). Understandably, however, many doctors have been reluctant to ask their patients to enter a study where whether they have their breast removed or not is decided at random, and many women are reluctant to do so. Obtaining the necessary 'informed consent' from them is difficult in such circumstances, particularly when such trials receive widespread publicity. Whilst some women are prepared to do this, there is an additional worry in some quarters that such women may not be representative of the population at large.

Despite these difficulties, many trials have been reported in the past 10 years which have greatly influenced the move towards as conservative surgery as is possible in each individual case.

Some early trials of conservative surgery as a routine procedure in early breast cancer showed that results were as good as those obtained with mastectomy, while others suggested a higher risk of local recurrence. In retrospect, this was probably because too low a dose of postoperative radiotherapy was used. An overview of results from a number of trials worldwide now shows that in appropriate cases (generally, women with early breast cancer, whose tumours measure less than 4 cm in diameter), conservative surgery followed by local radiotherapy can give as good results as mastectomy in controlling local disease and preventing recurrences, and yields similar survival figures.

The overall conclusion is that, provided the procedure is accompanied by whatever form of radiotherapy or other additional treatment is judged necessary, neither survival nor local recurrence is compromised by conserving the breast wherever possible.

Choice of operation

It can now be said with confidence that attempts to preserve a woman's breast do not pose a hazard to her life. Furthermore, the more extensive the surgery the more difficult it is likely to be to recover from the operation.

In particular, lymphoedema of the arm is highest after axillary surgery plus postoperative radiotherapy to the armpit and lowest after conservative surgery alone.

Most surgeons nowadays do everything they can to leave a woman with her breast where possible, or to choose the least extensive operation compatible with the extent of her disease. The treatment choice will also be dictated by the need for postoperative radiotherapy to reduce the incidence of recurrence, and the acceptability of this to the woman. Since the choice of operation is only partly on medical grounds, it is an appropriate development that the decision is generally now made in consultation with the woman and her family.

Lumpectomy is generally offered to women with early breast cancer, which in broad terms means small tumours with no evidence of spread. Nevertheless, as discussed above, not all patients are suitable for breast conservation, and the cosmetic results may be disastrous if it is attempted in inappropriate cases. However, radical mastectomy is no longer used routinely because its mutilating effects are not compensated by any increase in survival rates, and a majority of women for whom mastectomy is recommended will have either simple or Patey mastectomy. For both of these a cosmetically acceptable appearance can be achieved when dressed, even in low necklines, by the use of modern false breasts (prostheses). In addition, reconstructive surgery is increasingly available to restore both a breast mound and a nipple if a woman wishes this.

6

Recovering after surgery

Rehabilitation after treatment of breast cancer encompasses both physical and psychological adjustment. The latter is dealt with in Chapter 10. Physical considerations include dealing with the after-effects of surgery, the use of prostheses (false breasts), and reconstructive surgery.

Arm and shoulder problems after mastectomy

Arm exercises

The classical radical mastectomy was notorious for causing a stiff arm because of the muscles removed, but as this operation is now rarely used, severe arm and shoulder problems have largely disappeared. Nevertheless, women undergoing breast or axillary surgery may suffer minor damage to the nerves supplying the muscles, and still need to exercise to ensure they retain full use and flexibility of their arm and shoulder muscles.

A comprehensive range of exercises developed for mastectomy patients should be commenced as soon as possible after surgery. They will not damage the wound or break the stitches. Usually a physiotherapist will be on hand to teach and supervise the exercises while a woman is in hospital, so that by the time she goes home she should be able to brush the back of her hair and do up zips on the back of clothing. Arm exercises may also reduce both the risk and severity of lymphoedema.

1

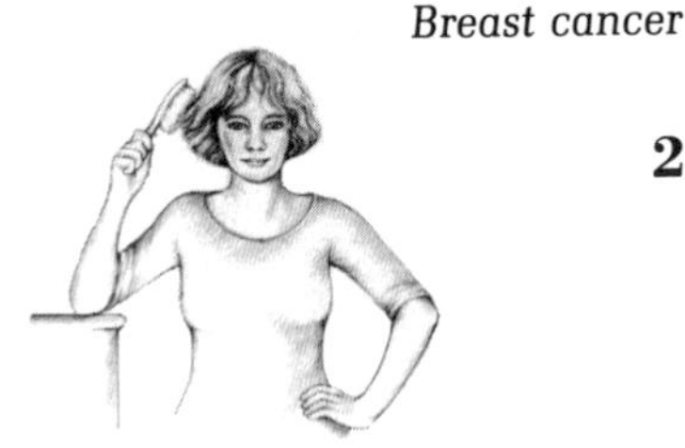

Hair-brushing exercise
(for hospital)
Rest elbow on bed-table. Keep head erect. Start by brushing one side only, then gradually increase to whole head. Don't overdo, but be persistent.

2

Squeezing and relaxing hand
(for hospital)
A rubber ball or similar object may be used.

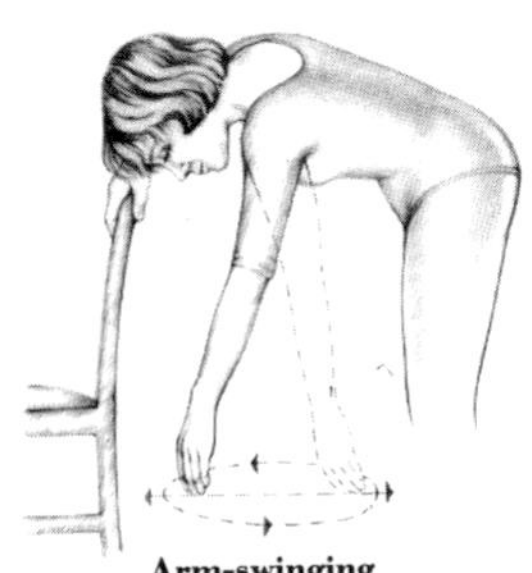

Arm-swinging
Place unaffected arm on back of chair and rest forehead on arm. Allow your other arm to hang loosely and swing from shoulder, forwards and backwards, then side to side and in small circles. As arm relaxes, increase length of swings and size of circles. Swing until arm is relaxed.

4

Bra-fastening
Extend arms, drop hands from elbows, then slowly reach behind back to bra level.

Wall-reaching
Feet apart for balance. Stand close to and facing wall. Start with hands at shoulder level and gradually work hands up the wall. Slide hands back to shoulder level before starting exercise again. Do slowly several times a day. Mark spot reached and aim higher each time.

Arm exercises after breast surgery.

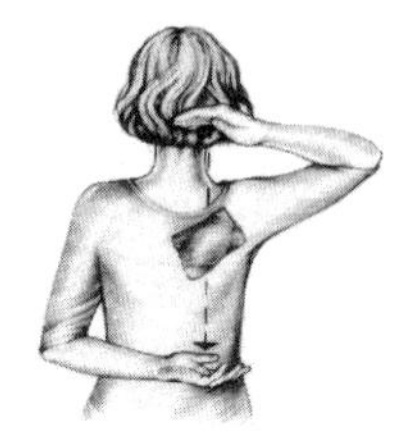

Bean-bag exercise

(A small purse or cosmetic bag will do just as well)

Drop bag from right hand over right shoulder into left hand at back. Repeat five times and do with opposite side.

7

Rope-pulley exercise

Throw rope or dressing gown cord over top of open door. Sit with door between legs. Hold lower end in hand on the side of your surgery and gently pull other end. Raise arm as high as possible each time, until full elevation.

8

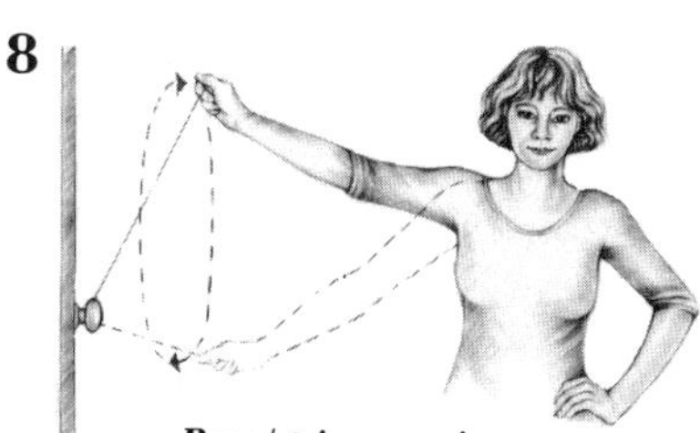

Rope/string exercise

Attach rope to doorknob or handle. Make small circles with rope moving entire arm from the shoulder. Do five times in one direction and five times in the other and gradually increase size of circle (by moving in closer) and number of circles.

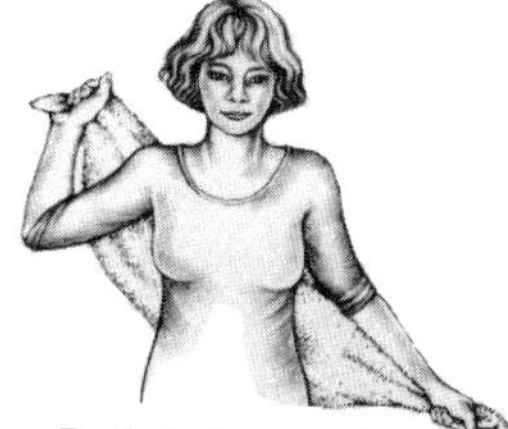

Back-drying exercise

With towel or similar item use a gentle back-drying motion. Reverse procedure.

Exercises should be continued when a woman gets home. A certain amount of discomfort is to be expected, but she should not overstrain herself so that movements cause severe pain. The best rule is 'little and often', with the exercises done daily in several short periods rather than prolonged sessions.

Lymphoedema

A swollen, aching arm due to blockage of the lymph circulation is also less often a problem since the move towards less extreme surgery for breast cancer. However, some women still suffer from this complication, especially if they have had the underarm lymph nodes removed or have had radiotherapy to this area.

It is important to protect the swollen arm from unnecessary knocks and scrapes, so gloves should be worn for rough household chores, gardening, or other heavy work. However, movement should not be unduly restricted. The exercises shown in the previous figure can actually help the swelling to go down, although some manoeuvres like changing gear when driving or hanging up washing may have to be reintroduced gradually, and it is probably best to use the other arm where possible for lifting heavy weights.

The arm should be raised up whenever possible, especially in bed at night when it can be propped up on pillows, and some women are advised to wear elastic bandaging on the arm overnight; these measures may help to reduce the swelling to tolerable levels.

Prostheses

After a mastectomy a woman is given a lightweight, temporary false breast (prosthesis) to wear for the first few weeks while the scar is still healing. She can then be fitted with a permanent prosthesis made from moulded silicone which fits in her bra.

Breast prostheses come in a huge variety of shapes and sizes

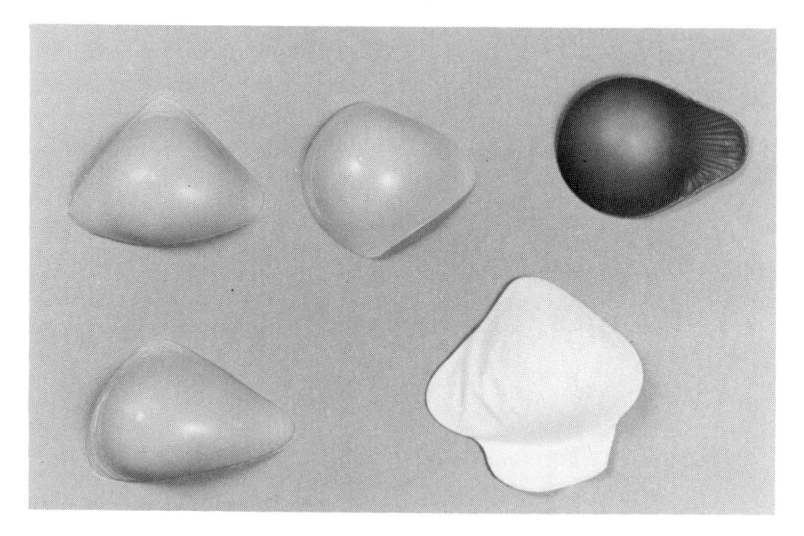

The choice of modern breast prostheses.

and every woman who has had a mastectomy is entitled to have whatever prostheses her surgeon judges suitable prescribed free under the NHS. Many hospitals now have a nurse or physiotherapist specially trained in the use and fitting of breast prostheses who can advise women on how to wear them in comfort. With simple or modified radical mastectomies a prosthesis gives an excellent outward appearance even with a fairly steep cleavage or while wearing a swimsuit. Small ones can also be supplied for women with uneven breasts after conservative surgery.

Breast reconstruction

Although more than half of all women with breast cancer no longer have a mastectomy, not all tumours are suitable for conservative surgery. In addition, a small but significant proportion of women actually choose mastectomy over other options. There are also women who have had a mastectomy in the past who remain unhappy with their appearance, or who did not know that anything could be done to restore the appearance of their lost breast. A few women who have conservative surgery lose a sufficient amount of breast tissue to create an obvious hollow in the breast or discrepancy with the other side.

Reconstructive surgery is the obvious answer for such women, and except for the very unfit most women are suitable for some kind of reconstructive procedure. Even women known at the time of diagnosis to have widespread disease may still have a life expectancy of several years, and reconstruction can dramatically improve their quality of life during this time.

In recent years the techniques have been much refined, and the aim is no longer simply to 'fill the bra', but to produce a breast which looks attractive and matches the other side. A cosmetic result which pleases the woman can now be achieved in most cases.

Until recently this option was only available to a few women motivated to seek out a plastic surgeon to treat them, sometimes privately at great expense (the cost is about £3000). Reconstructive surgery is still rarely offered as a routine procedure, and not all treatment centres have access to a surgeon experienced in reconstruction. Nevertheless, it is becoming increasingly popular and is available on the NHS.

Whether a woman is offered reconstruction or not will depend on the attitude of the surgeons in the hospital treating her, although every woman has the right to ask her GP for referral to a different surgeon if the one who originally treated her will not agree to reconstruction. Some surgeons offer reconstruction at the same time as the mastectomy, while some centres are not able to do this, and some surgeons prefer to delay until later. If there has to be a gap, many women no longer wish to face a second operation.

Although, overall, less than 5 per cent of women having a mastectomy have breast reconstruction, in centres where reconstruction is offered to all mastectomy patients, about half choose to have it.

Effects of reconstruction

Mastectomy represents a deformity and a threat to a woman's body image and sense of femininity, as well as possibly changing her lifestyle and the way she dresses. It also serves as a constant reminder of the cancer.

Reconstruction abolishes the need for prostheses, special clothing or restricted dress styles or activities, and can help to increase a woman's sense of self-esteem and optimism.

Reconstruction does not affect the cancer or the options for treating it. It makes no difference to the outcome of cancer surgery or of radiotherapy, chemotherapy, or hormone treatments, nor does it make it more difficult to follow-up a woman or detect any recurrence of the tumour.

Immediate or delayed reconstruction?

Breast reconstruction may be performed at the time of the original mastectomy, or at any time afterwards. Immediate reconstruction means that the woman awakes from her operation with a breast still present, even though it is different to the one she started with. This obviously results in the least psychological trauma, and avoids having to face further surgery at a later date. A very few treatment centres offer immediate reconstruction, and find that more women take up the option when they only have to have one operation.

However, even centres which do offer reconstruction to most mastectomy patients are seldom able to perform the reconstruction during the same operation as the mastectomy, since this usually requires two different specialists to be present, a cancer surgeon and a plastic surgeon, so it becomes a much more complicated operation both to arrange and to perform.

Some surgeons advocate a delay, either in case of early tumour recurrence which could require further treatment of the local area, or because they feel that since the results are never a perfect match with the other breast, a woman will be less satisfied with the new breast after immediate reconstruction.

In practice most reconstruction patients have to wait for a year or more after initial surgery. Some women find out about reconstruction many years after their mastectomy, and there is no reason why it cannot be carried out even decades later.

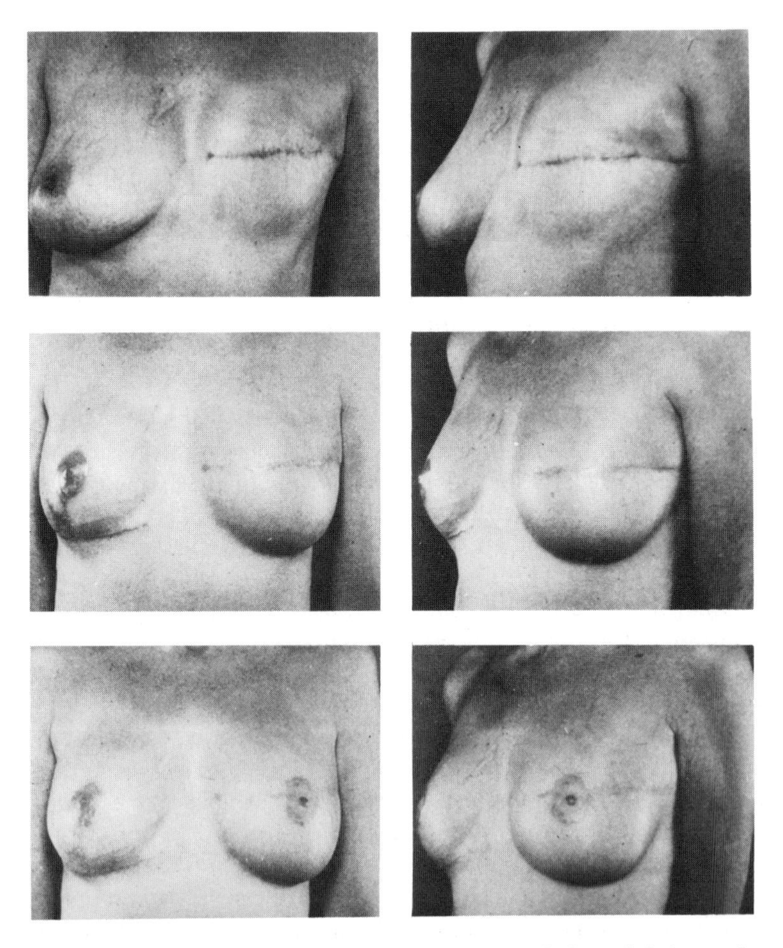

Left Becker reconstruction – breast being expanded, with left nipple reconstruction and simultaneous right breast reduction.

Techniques of reconstruction

Silicone gel implants

The simplest type of reconstruction involves a silicone implant inserted under the pectoralis major muscle of the chest wall. The implant is a soft bag filled with silicone gel which can be put in either at the same time as the mastectomy or at a later stage, usually through the same scar. An implant simply replaces the tumour and surrounding tissue which has been taken out and recreates the breast mound.

The full effect will not be seen until the skin and surrounding tissues have had time to stretch to accommodate it, which takes several months (just like breast enlargement in pregnancy). Although different sizes are available to match the other breast as best as possible, size is limited by the amount of skin left after mastectomy, so implants alone are only suitable for quite small breasts, an A or B cup. They also sit quite high on the chest, so the size and position of the other breast need careful assessment to avoid a lopsided appearance.

Most women have no problem with this form of reconstruction. However, the body reacts to a silicone implant by forming a capsule around it, and sometimes this can contract, becoming firm and leading to a hard feel of the breast. These implants are the same as those used in breast enlargement operations, and there has been much concern and media excitement generated by reports from the USA that in some cases they may leak and lead to autoimmune diseases such as arthritis, rashes, and allergies, and even to cancer.

Most assessments have not borne these fears out and there is certainly no evidence that they can cause cancer. In the UK experts remain unconvinced that any risk exists, but just to be sure the Department of Health has set up a breast implant register to monitor all women having implant surgery, whether after mastectomy or for cosmetic reasons. Early reports of implants 'exploding' and collapsing within the breast when women who had had breast enlargement operations went flying in pressurized aircraft seem to have been anecdotal and certainly this is not a recognized complication today.

Tissue expanders

A slightly more complicated option, particularly suitable for women with larger breasts, is to insert a type of 'balloon', instead of a silicone implant. The bag is then gradually inflated with salt water (saline) which is injected through a valve implanted under the skin, usually at the woman's side. The injection port is a small plastic box which should not bother the woman at all while it is in place, and the process of injecting fluid hurts no more than the fluid injected during an immunization.

About 50 ml of saline are injected every week or two until the new breast mound is somewhat larger than the opposite breast. This enables gradual stretching of the skin and surrounding tissues, and leads to a more natural breast shape and usually a better cosmetic result.

The injection procedure continues until the breast containing the balloon reaches about one and a half times the size of the other breast, so that when the fluid is removed the breast will hang down in a natural way. The bag may then be removed

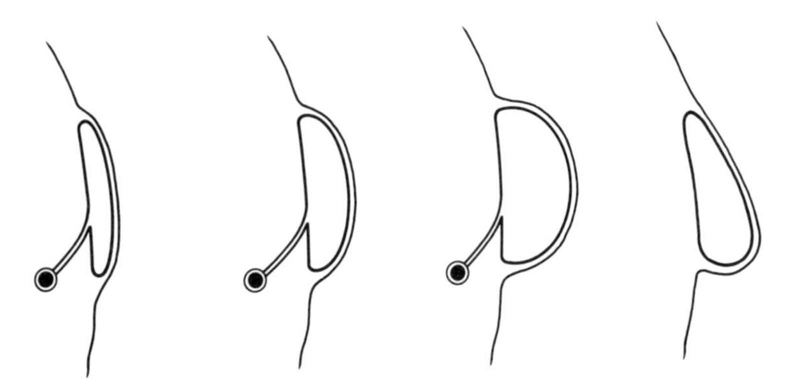

'Becker' reconstruction of the breast.

and replaced with a permanent silicone implant at a second operation.

Some newer devices such as the 'Becker' prosthesis have a silicone implant already incorporated, so only the valve needs to be removed and this can be done under local anaesthetic (see figures on pp. 68 and 70).

Musculocutaneous flaps

With this technique a flap of skin and muscle is taken from somewhere else on the body, usually the abdomen or the back, and used to reconstruct the breast shape, often with an implant as well. This can be useful if the amount of tissue

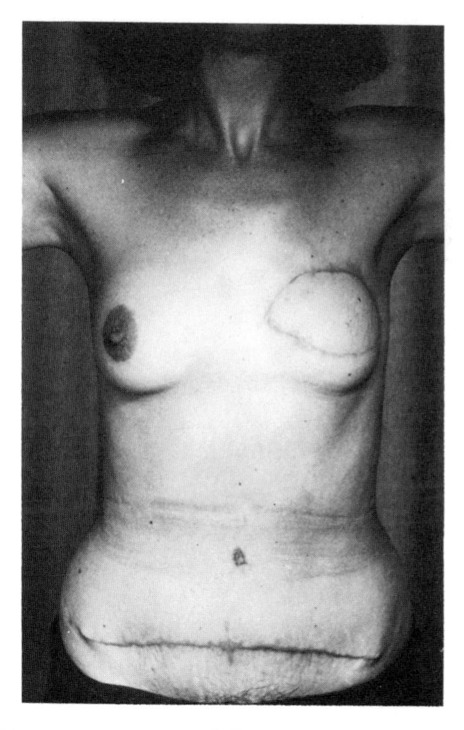

'TRAM' (Transverse Rectus Abdominus Musculo-cutaneous) flap.

removed with the tumour was very large or if the woman has naturally large breasts. It can also be used if for some reason the skin is unlikely to stretch to accommodate an implant, as can sometimes happen after radiotherapy to the breast area.

A flap is usually swivelled around from an adjacent area to the breast along with its own blood supply. A new technique enables free flaps to be moved and connected to a new blood supply on the breast, although this is a much more complicated procedure.

The main disadvantage of flaps is the need to damage an additional area of the body and the significant additional scar created at the site where the flap is taken. The most common flap is called the latissimus dorsi flap (after the muscle it uses),

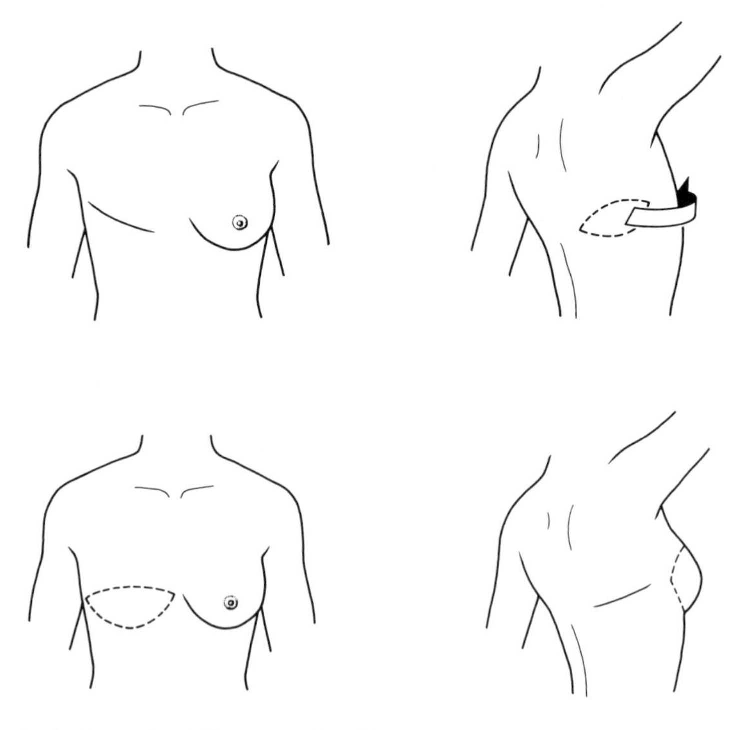

Latissimus dorsi flap reconstruction.

and a newer flap based on abdominal muscle (the 'TRAM' flap) is also becoming popular. This is illustrated in the figure on p. 71.

Nipple reconstruction

Some women are happy with just the shape and mound of a new breast to fill out bras and swimsuits and ensure a normal looking cleavage no matter how plunging the neckline. Others use plastic stick-on nipples to give the right contour to the breast.

If the woman wants it, however, surgical nipple reconstruction can give a reasonable cosmetic appearance, and more and more women having reconstruction operations now want this to give an appearance as near as possible to a normal breast. However, it does entail yet another operation, since nipple reconstruction is best done after an interval of several months to allow the reconstructed breast to settle, so that the new nipple ends up at the right height to match the other side.

If it is large enough, tissue can be taken from the other nipple, if not it is taken from the inner thigh or the labia. The colour of the new nipple can be darkened by tattooing if necessary, and if a projecting nipple is required to match the other side a small amount of cartilage from an earlobe can be used.

The opposite breast

Sometimes to achieve a matching result it is necessary to operate on the opposite breast, either to reduce its size (reduction mammoplasty) or to lift it up (mastopexy).

7

Additional treatments for breast cancer

Most women with breast cancer have some form of surgery to remove the tumour, either by taking out the lump alone or by removing the whole breast. Even so, there is a risk that some cancer cells may be left behind because often, by the time breast cancer is diagnosed, some of the cells have spread beyond the lump itself.

It is believed that even where cancer appears to involve only a single site within one breast, in 40 per cent or more cases there may be multiple other areas of change within the same breast which are either premalignant or already cancerous. In addition, it is obvious that even when the spread of a tumour is not detectable at the time of diagnosis, in many cases it must have already occurred because secondaries appear later on. It is believed that small deposits of cancer cells (micrometastases) have spread in the body in up to 70 per cent of cases by the time the cancer first becomes apparent.

If these cells are allowed to grow, cancer could come back in the same site (recurrence), or could spread to form secondary tumours elsewhere in the body. So additional forms of treatment (adjuvant therapy) are often given as an extra insurance to try to destroy any remaining cancer cells.

The ideal anti-cancer therapy would kill off cancer cells without harming other, healthy cells in the body. Unfortunately, there is as yet no known treatment which can do this completely, so doctors have to rely on methods which target particular features of cancers. This means that although the cancer cells are the most likely ones to be attacked, some healthy cells may be affected as well, leading to a variety of side-effects.

At present adjuvant treatment may take the form of radiotherapy, given to the breast area to reduce the risk of recurrence, or hormonal treatments or anticancer drugs (chemotherapy) aimed at reducing the risk of secondary growth elsewhere in the body (systemic treatment). Because of the risk of micrometastases, some form of systemic treatment may be advisable for most patients, even after surgery and radiotherapy combined.

As these treatments have side-effects, each woman and her doctor must carefully consider the balance of risks versus benefits in her individual case before deciding whether to use them or not. Where adjuvant therapy is thought to be worthwhile, it has been shown to improve a woman's chance of survival by about 25 per cent.

Radiotherapy

Radiotherapy, or radiation treatment, is often given after surgery for breast cancer to try to ensure that all the cancer cells in the local area of the affected breast have been destroyed. It uses a type of X-ray carefully targeted to the breast area and chest wall, and occasionally to the underarm region and area above the collar-bone.

A course of radiotherapy usually involves up to five treatment sessions a week over about six weeks, given as an out-patient. Each dose of radiation is carefully calculated and takes several minutes, during which the woman is asked to lie very still, but otherwise the experience is no different to that of having an ordinary X-ray, and does not hurt.

Initially a careful assessment is made of the area to be treated, which is marked with a pen and a tiny tattoo and women are generally asked not to wash the area during the course of treatment, to avoid irritating the skin.

Because they are working with X-rays all the time, the staff who monitor the treatment do so from outside the treatment room, to protect them from too many doses of radiation. But radiation treatment does not make the patient radioactive and there is no danger to a woman's family, including children, from coming into contact with her.

The side-effects of radiotherapy have often been exaggerated in the past, and serious side-effects are much less common with modern techniques. Radiotherapy to the breast area does not cause infertility, nor will a woman's hair fall out.

The need for regular, frequent journeys to hospital is a major drawback, and as well as the inconvenience some patients find this, and the treatments themselves, very tiring. It is often a good idea to plan for a rest each day on getting back from the hospital, and to keep daily routines as uncluttered as possible.

Very occasionally some women feel sick or queasy for a while after each treatment and may need to plan meals carefully or to take antisickness tablets.

Some women, but not all, find that radiotherapy makes their skin sensitive, rather like being in the sun for too long. The skin in the exposed area may become darker, slightly itchy or sore, and sometimes small blood vessels in the skin may burst, leaving tiny red marks. Like sunburn, those most at risk from skin problems are those with red hair and fair skin.

Very rarely, the top part of the lung may be affected by the radiation and become stiff and fibrous (radiation fibrosis), which can cause a dry cough or breathlessness many months afterwards.

Radiotherapy is not generally given after mastectomy unless the muscles of the chest wall are at risk from local recurrence, to avoid painful arm swelling (lymphoedema) which may otherwise result from poor drainage of lymph fluids. This is therefore an increasingly rare complication.

Radiotherapy can interfere with the body's immune system, which is important not only in defence against infection but also in controlling the spread of cancer cells. Doctors used to worry that this might increase the risk of cancer coming back after treatment, but long-term studies of women who have had radiotherapy have not borne this out.

Usually all side-effects of radiotherapy disappear once the course of treatment is ended or within a few weeks afterwards, but it is a good idea to avoid sunbathing for a year or so after radiation treatment.

Doctors are still unsure which women should have radiotherapy after breast cancer surgery. After conservative surgery, it undoubtedly reduces the risk of the cancer returning in the same area. However, only about 30 per cent of such women are at risk of this anyway, so treating all patients would mean many women having radiotherapy unnecessarily. Furthermore, while it can be devastating for women with small tumours to get a local recurrence of cancer some years after apparently successful treatment, such recurrence does not seem to affect long-term survival provided radiotherapy is given at this stage.

To try to resolve the question of which women should have radiotherapy with their initial surgical treatment, a national trial is now being carried out to assess which women gain the most benefit from it. Meanwhile there is no doubt that radiotherapy should be given to women at the highest risk of their cancer recurring, such as those whose tumours were found to be particularly large or aggressive. For women at lower risk, it may be possible to monitor closely and use radiotherapy only for those women whose cancer does come back, when treatment at an early stage can still cure it without compromising survival.

A special type of radiation treatment now coming back into fashion and sometimes offered to women after conservative surgery uses fine wires or needles containing a radioactive source to apply high-dose treatment to the surrounding area from within the breast itself. The wires are put in under general anaesthetic and left in place for several days, when they may feel uncomfortable but should not hurt, after which they can usually be removed without another anaesthetic. Because this type of treatment does give out radiation, the woman will be asked to stay in hospital, usually in a side ward, and while most people are safe for short visits, children and pregnant women are advised to stay away until the wires are taken out.

Most doctors nowadays will spend time discussing the risks

and benefits of radiotherapy after surgery so that each woman can make an informed choice in her own particular case.

Adjuvant systemic treatment

This means treatment aimed at destroying any cancer cells which may have escaped from the breast and lodged in other parts of the body. Systemic treatment affects the whole body and not just the breast. There are two main types, hormone treatment and anti-cancer drugs or chemotherapy.

Hormone treatment

Breast cancer development is influenced by the levels and fluctuations in a woman's hormones, and changing these levels can help in the treatment of some forms of breast cancer.

Because some breast cancers are very sensitive to the levels of oestrogen, the 'female' hormone, various forms of treatment have attempted to reduce or abolish production of oestrogen within a woman's body. These have included surgical removal of the ovaries or their destruction by radiation (ovarian ablation), treatment of the adrenal glands (which sit just above the kidneys) with drugs to stop the small amount of oestrogen produced there, and anti-oestrogen drugs such as tamoxifen.

Ovarian ablation

Abolishing the secretion of oestrogen from the ovaries, either by surgery or radiotherapy, has been shown to increase overall survival in premenopausal women with breast cancer by about 30 per cent. This treatment also reduces the number of women who get recurrent disease by about 25 per cent. However, most specialists prefer not to use these methods in younger women because they cause infertility and bring on an early menopause.

Goserelin

An alternative approach now being tested is to use a new drug called goserelin (Zoladex), which is given by injection. It turns off the production of oestrogen from the ovaries by disrupting other hormones coming from the brain which control their output. The reduced oestrogen levels may cause menopausal symptoms, but the effect is reversible when the drug is stopped.

Tamoxifen

Another option is a drug called tamoxifen, which blocks the stimulatory effect of oestrogen on breast cancer cells. It may also have other actions, for example in stimulating the body's own natural anti-cancer defences.

Over the last ten years tamoxifen has produced a modest but extremely exciting breakthrough in breast cancer treatment. A 20 mg tablet, taken once a day for between two and five years, will give a 20 per cent to 30 per cent reduction in the risk of dying from breast cancer. The benefits appear to continue for at least ten years.

Tamoxifen has very few side-effects, although some premenopausal women may experience menopausal symptoms such as hot flushes. On the plus side, tamoxifen has actions similar to natural oestrogen on other body tissues, so it reduces the levels of fats in the blood (and thus reduces the risk of heart disease) and prevents the bone loss (osteoporosis) which occurs in postmenopausal women.

The benefits of tamoxifen apply to all women irrespective of their age or the stage of their cancer, but women over the age of 50 seem to gain most benefit. Women with oestrogen receptor-positive tumours are more likely to respond to tamoxifen, although some benefit is seen in those with receptor-negative tumours as well.

Most postmenopausal women are therefore given tamoxifen after breast cancer surgery, irrespective of their receptor status or lymph node involvement. Hopes that tamoxifen alone might be suitable to treat elderly women, for whom the hazards of surgery are greater, have not been borne out. Recently a large trial in women over 70 showed that tamoxifen alone produced an unacceptably high recurrence rate, leading to large numbers of these women having to have surgery later on to control the tumour after tamoxifen alone had failed to do so.

The role of tamoxifen in premenopausal women is less certain, and the optimum adjuvant treatment for younger women remains unclear. Tamoxifen may threaten a woman's fertility, particularly as she nears the menopause. It may also lead to irregular periods or an early menopause.

Tamoxifen may also play an increasingly important role in prevention of breast cancer, which is discussed in Chapter 11.

Anti-cancer drugs

Anti-cancer drugs (chemotherapy) aim to track down and kill cancer cells anywhere in the body. They are often given after surgery for breast cancer because of the high risk that a few cancer cells may have spread outside the breast even before the tumour was diagnosed.

Chemotherapy is most likely to be recommended for premenopausal women whose underarm lymph nodes are involved or whose tumour is particularly aggressive. In this group it can yield up to a 40 per cent decrease in the risk of dying of the disease. It may also be offered to postmenopausal women with involved lymph nodes, although generally older women fare better on tamoxifen, which has fewer side-effects.

For a few women who do not want surgery or whose tumours are judged unsuitable for surgery, chemotherapy may be given as the first or only treatment. It may enable some cancers to shrink to a size where they become operable by wide local excision instead of mastectomy, or even to disappear altogether.

Many different kinds of anti-cancer drugs are available and they are given in different combinations, so their effects and side-effects vary greatly from woman to woman. They are usually given in cycles at about monthly intervals for six months. The drugs are usually given by injection or through a drip inserted into a vein in the arm, often with tablets to take as well. The drugs are sometimes given as an out-patient treatment but may mean staying in hospital overnight after each treatment for its effects to be monitored.

Although anti-cancer drugs are designed to be more damaging to cancer cells than to normal ones, they may also damage healthy cells (so are sometimes called cytotoxic drugs), particularly those which have a rapid growth rate. So each course is usually followed by a few weeks' break for the body to recover. Usually a blood test will be taken before each treatment to check that levels of white blood cells, important in the body's defences against infection, have not been too affected.

If the number of cells is too low, the next course of treatment may be delayed until it returns to a safe level. Sometimes treatment with antibiotics or a blood transfusion are needed during a chemotherapy course.

Other side-effects may include tiredness, nausea, hair loss, mouth ulcers, loss of appetite, and diarrhoea, a combination which can make a woman feel very unwell and miserable. However, many unwanted symptoms can be helped by simple treatments, such as mouthwashes to help combat mouth soreness, and they do disappear rapidly once treatment is finished. Powerful anti-emetic drugs have been specially developed to combat the nausea and sickness associated with anti-cancer drugs, and treatment staff will advise on these if necessary.

It is uncommon for anti-cancer drugs to cause complete hair loss, although many women find their hair thins a little during treatment. Nevertheless, hair loss is for some women the most distressing prospect, especially coming on top of the traumas they have faced with breast cancer surgery. If hair loss is very noticeable, wigs are available on the NHS, and the hair will grow back after treatment, or even before the treatment is finished, although it may be slightly more curly and sometimes the colour changes a little.

Menstruation is often disrupted during chemotherapy, in some cases with periods stopping altogether, and fertility may be affected by some types of anti-cancer drugs. About 40 per cent of women on chemotherapy will become infertile, so if a woman wants to have children in future this will need to be carefully discussed with her doctor before treatment begins. Younger women tend to continue their periods or restart them after treatment more often than women closer to the menopause.

Overall, treatment with chemotherapy yields a 28 per cent reduction in the annual rate of recurrence and a 16 per cent reduction in the yearly death rate. However, because anti-cancer drugs are relatively toxic, they may be best reserved for women most likely to benefit. Studies suggest that modern treatment with multiple drugs (cyclical combination chemotherapy) is particularly suitable for premenopausal women whose cancers have been shown to involve the lymph nodes. In this group, it can reduce the chance of dying over the next ten years by 30 per cent to 40 per cent. For women who have already had the menopause, hormone treatment with tamoxifen (see previous text) may be more appropriate, although some specialists advocate chemotherapy for postmenopausal women as well.

Most chemotherapy courses are given after the woman has had some kind of surgery. However, recent research suggests that chemotherapy given before surgery may give an even better increase in survival. Trials comparing pre- and postoperative chemotherapy are now underway and if this approach proves favourable it may become more common in future.

Finally, it could be that using hormone treatment and anti-cancer drugs together could produce even better results than either treatment alone, and this possibility is now being tested in clinical trials.

'Complementary' medicine

In discussing non-surgical treatments we should also mention the profusion of therapies offered outside the framework of orthodox medicine, some of which claim to be able to help cancer patients.

The dilemma about what to call these illustrates in part

why many doctors feel somewhat dubious about them. Some branches of so-called 'alternative' medicine may recommend that women follow their treatments *in preference to* orthodox medical treatment. Doctors have seen cases of women who have done this and thereby allowed their cancer to grow to an advanced stage untreatable by orthodox methods.

Various therapies which could more reasonably claim to be 'complementary' or *additional to* orthodox medicine may have more to offer in that they do not deny patients the benefits of conventional treatment. This is not the place for a discussion of the details or the merits of individual therapies, many of which centre on diet, massage, relaxation, visualization, and meditation techniques which are not necessarily specific to cancer treatment.

There is no doubt that some complementary therapies make women feel better in themselves, and that many techniques can aid psychological adjustment and enhance enjoyment of life. On the other hand, treatments involving strict diets may make women feel worse, and if they have not been followed to the letter any recurrence can provoke needless guilt and self-blame. In addition, some therapies are based on an underlying philosophy that disease stems from a person's attitudes and ways of dealing with stress, which can make people feel as if the tumour is their own fault.

Generally complementary therapies claim to be holistic treatments, meaning that they treat the patient as a whole person, with bodily, mental, and spiritual attributes, and not just a collection of symptoms around a tumour. This is part of their growing appeal. However, it is what good orthodox medicine should do as well, and while some criticisms are well-founded, complementary therapies do not have a monopoly on the holistic approach.

Perhaps the critical test for any woman contemplating complementary therapy is whether it can be integrated with orthodox treatment. If her doctor agrees that it can do no harm, and provided the therapist does not attempt to dissuade her from having proper medical treatment, then what matters is whether or not it makes her feel better.

8

Course and outcome

The fundamental question any woman with breast cancer is likely to want to ask is 'Will I survive?' It is extremely difficult for a doctor to offer her an accurate prediction of either her likely chance of cure or her potential lifespan, since at the time of diagnosis the disease pattern can vary enormously.

Some women have a slowly progressive form of breast cancer which can grow unnoticed for a while, eventually produce symptoms which lead the woman to her doctor perhaps years after it first began, and finally many years or even decades later lead to effects which may be fatal. Since the average age of onset of breast cancer is 55, a number of women will die of other causes even though they still have the breast cancer, whether recognized and treated or not. Other women can see a doctor for the first time with symptoms which indicate a cancer which has already spread far in the body, yet the primary tumour in the breast can be virtually undetectable.

The outlook and the chance of treatment succeeding will be affected by the grade and stage of the disease, by whether a woman has involved lymph nodes or not, by her general health and medical history, and perhaps by other influences such as whether her tumour is oestrogen receptor-positive or not. All these factors need to be taken into account, along with the treatment a woman receives, in assessing the likely outcome, or prognosis, of breast cancer.

Course of breast cancer if untreated

The results of any form of treatment for breast cancer need to be compared with the course it would take if left untreated. This is because, contrary to popular belief, cancer (of various kinds) is not invariably fatal, even without treatment. Cancers vary enormously in the length of time they take to cause symptoms, to become serious, or to lead to terminal disease.

We need to know how a woman would fare with no treatment at all in order to assess how much benefit is likely from the treatment on offer. The natural history of a disease also enables us to understand more about the stage of disease at presentation.

Medical knowledge of the full natural history of untreated breast cancer is incomplete. Although, unfortunately, some women still delay seeing a doctor until their tumour is quite advanced, even extreme cases will be offered some form of treatment. Our understanding will therefore probably always be less than perfect since it relies in large measure on historical information from the days when large numbers of women went untreated.

Historical records of breast cancer

Cancer of the breast was described by the ancient Egyptians and ancient Greeks, and mastectomy was probably practised in ancient Rome. Even in those days, physicians distinguished between early and advanced disease and noted that surgery was futile for the latter.

In the early eighteenth century it was recognized that breast cancer could spread to the lymph nodes in the armpit and that this implied a poor outlook. There are a number of reliable reports of the progression of untreated breast cancer from the nineteenth century. These suggest that on average infiltration of the skin by a breast cancer occurred 14 months after the tumour was first detected, ulceration of the skin occurred six months later, and fixation of the tumour to the chest wall after a further two months. In those with only a breast tumour detectable at the time of diagnosis, enlarged lymph nodes in the armpit appeared on average 15 months later.

Among women who survived long enough, the second breast was typically invaded by the cancer about three years after the lump appeared in the first one. One quarter of the women developed secondary tumours within a year, but another 25 per cent had not developed secondaries at three years, and after 10 years about 10 per cent of women had survived without any treatment.

More recently, the records of 250 women who died of breast cancer in the Middlesex Hospital cancer ward between 1905 and 1933 were assessed. Nearly all were admitted with either locally advanced or widespread disease, and the time from the recorded onset of symptoms to death was, on average, just over three years, with 18 per cent surviving for five years and 3.6 per cent for ten years.

It must be appreciated that in the past women probably did not go to a doctor until their disease was generally much more advanced than the small lumps mostly seen nowadays, and many probably did not go at all. Treatment would probably have been offered mainly to women judged to have less advanced disease, so untreated patients who were observed in this way were thus likely to be a selected group of the worst cases.

The importance of this distinction is shown by comparison with a relatively modern study from the Ontario Cancer Clinics published in 1965. Fifty women were identified (among nearly 10 000 cases) who had presented with early breast cancer and refused treatment, for a variety of reasons. About 70 per cent of this group survived for five years from the time of the first symptom, an outcome which does not compare unfavourably with that of many treated patients.

How is the course of disease affected by treatment?

Since breast cancer seems to fall into at least two categories, aggressive and slowly progressive disease, how can treatment be assessed? In particular, how—and when—can a woman be judged to be cured after treatment?

Surgery for breast cancer

Removal of the breast in the attempt to cure breast cancer is mentioned in the texts of several ancient civilizations, and has been variously attempted over the intervening centuries by many prominent Western surgeons. There are anecdotal reports of women in the early eighteenth century submitting to what must have been very primitive surgery in those days and surviving for decades, but these were undoubtedly the exception rather than the rule.

Mastectomy was attempted more often in the nineteenth century, when records show that nearly one in five women died from the operation, and four in five of the survivors developed recurrent disease. This seems to have been the case whether or not lymph nodes or muscles were removed as well as the breast tissue. Almost one in ten women lived for ten years after treatment, which suggests that surgery made little difference to their chance of survival.

The history of modern mastectomy may be said to date from the introduction of classical radical mastectomy, also called Halsted mastectomy after its originator, a US surgeon who operated on nearly 900 patients from 1889 to 1931. Six per cent died at or soon after the operation, and only 12 per cent survived for 10 years, but the frequency of local recurrence was reduced to 30 per cent. Thus the introduction of the radical procedure markedly improved local recurrence but made little difference to long-term survival.

The real breakthrough came in the 1940s with methods of staging the progression of the disease. This led to better selection of women who were most likely to benefit from aggressive surgery, and hence to an increase in the proportion who lived for 10 years (the 10-year survival rate) after mastectomy from about 10 per cent in the 1920s to about 50 per cent in the 1950s.

The current position seems to be that even the most radical surgery cannot, by itself, increase the average survival rate much more than this. Despite all the supposed benefits of earlier diagnosis, advances in surgical techniques, and postoperative radiotherapy, a woman's chance of living for 10 years after breast cancer surgery remained stubbornly stuck at no better than an even 50–50 for many years. Only recently has the use of effective adjuvant treatments produced a marked improvement.

Non-surgical treatments

The effect of the various adjuvant treatments on the likelihood of survival and of disease recurrence is discussed in Chapter 7. In summary, it can be said that postoperative radiotherapy has little effect on survival rates but does reduce the risk of local recurrence after surgery, to which two or three women in every ten will otherwise succumb.

In general, the use of tamoxifen in postmenopausal women, or stopping ovarian oestrogen production in premenopausal women, can both increase average ten-year survival after breast cancer surgery by up to 30 per cent as well as reducing the annual recurrence rate by up to 25 per cent. Chemotherapy given to premenopausal women yields an overall 28 per cent reduction in the annual rate of recurrence and a 16 per cent reduction in death rate, with especial benefits for women with involved lymph nodes, for whom the risk of death within ten years may be reduced by 30 per cent to 40 per cent.

It is only in the last year or so that these benefits have been reported to make a difference, finally, to the overall outlook for breast cancer. After analysis of many trial results from around the globe, it seems that adding hormone treatment or chemotherapy to the initial treatment of breast cancer may improve average survival by about 25 per cent, with a similar improvement in the other main outcome measure, time free of recurrence.

Grade and stage of disease

The benefits yielded by the various forms of treatment discussed above are, of course, averages. For an individual woman the outlook will be affected by the characteristics of her own tumour and its stage of development. Aggressive breast cancers which spread at an early stage may be incurable by local treatment, whereas slow-growing tumours may be potentially curable by any form of surgery or radiotherapy to the local area.

As discussed previously, one of the most important factors affecting the outlook for a woman with breast cancer is whether or not the axillary lymph nodes are affected at the time of diagnosis. After surgical treatment alone, the relapse rate at five years is 20 to 30 per cent in women without involved nodes, compared with 60 per cent among those with involved axillary nodes.

The degree of involvement of these nodes gives further clues. Generally, the more nodes involved, the higher the chance of relapse. If four or more nodes are involved, or if those further away from the breast are affected, the outlook is poor. Even after surgical treatment, only 20 per cent of these women are likely to survive for ten years.

Since involved lymph nodes are not now believed to be an additional source of spread of the disease, but rather to indicate that spread has already taken place, whether they are removed or not seems to have little effect on the eventual outcome. Similarly, the outcome is now believed to be determined not by the extent of surgery to the breast but by the amount of undetected secondary spread present at the time of diagnosis.

Nevertheless, knowledge of the stage of a patient's disease can help doctors to decide on additional non-surgical treatment. For instance, women with 10 or more involved nodes are particularly likely to relapse, so it may be appropriate to offer them more intensive chemotherapy.

This demonstrates that improvements in outcome of treatment for breast cancer may come about not only as a result of better means of defeating the cancer, but also as a result of better selection of women for various types of treatment. The focus of much current research work is on attempts to predict more accurately the likely course of disease in an individual woman in order to target treatment to women most at risk.

This approach offers the chance both to improve the prognosis in women for whom there is a potential for cure, where the discomforts and hazards of treatment can yield a worthwhile benefit of increased years of active life, and to reduce the need for unnecessary treatment in women with aggressive or spreading disease, who may be better served by palliative treatment designed to relieve symptoms and enable them to spend their remaining time in comfort.

Follow-up of treated patients

For a woman who has had successful treatment of breast cancer, the next most important question is at what stage can she be considered 'cured', given all these variable factors?

The chance of survival increases the longer a woman lives after surgery without developing secondaries or recurrent disease, since women with aggressive disease which has not been caught by treatment are more likely to succumb early on. The proportional death rate is therefore greater in the earlier years, and women are usually closely monitored for at least five years after treatment to detect possible recurrences or secondary spread.

However, late recurrences are still possible and experience has shown that a much longer follow-up is needed before a woman can truly be assured that her disease will not come back. Today it is conventional to talk in terms of ten-year survival rates, although several studies have suggested that, after allowing for deaths from old age and other causes, women who have had breast cancer may still have a small excess risk of dying compared with women of the same age even 40 years after treatment.

Initial survival is heavily influenced by age, menopausal state, stage of disease, and node status. Broadly speaking, women who develop breast cancer under the age of 35 fare badly, as do the few whose disease starts around the time of the menopause. Premenopausal women aged 35 to 50 fare best, while postmenopausal women do slightly less well. However, after the first five years free of disease recurrence, these conventional factors seem of little value in predicting prognosis.

Recurrent disease

It is not uncommon for cancer to recur in the area of the treated breast, irrespective of the form of surgery and the use of adjuvant treatment. Recurrences occur in 20 to 30 per cent of women who have undergone surgery aimed at cure of the disease, and while most women quite reasonably dread the thought of the disease coming back, it need not be a terrifying outcome.

Radiotherapy for local recurrence can still be curative if given early, hence follow-up schedules for breast cancer patients are designed to enable detection of recurrent disease at a stage where delayed radiotherapy can produce adequate control in most cases. However, there does seem to be a hard core of 5 per cent to 10 per cent of cases in which recurrence eventually catches up with the patient and leads to spread of the disease.

Cancer of the opposite breast

One of the strongest risk factors for the development of cancer in one breast is having already had cancer in the other breast. After treatment of a primary breast cancer, even if apparently cured, a woman is at high risk of a new primary tumour developing in the opposite breast. With the exception of tamoxifen, which seems to reduce the risk, this is the case irrespective of adjuvant treatment being given.

The risk has been estimated at four to seven times that of developing a first tumour, but the situation remains uncertain because to an extent the evidence is confusing and figures vary according to the method of detection used and the characteristics of the abnormalities detected.

On the basis of symptoms, physical examination, and mammography of the opposite breast, about 5 per cent of women are found to have a primary tumour in both breasts at the same time, and around 7 per cent may develop a second primary tumour at a later stage. Overall the lifetime risk of a second primary tumour may be as high as 20 per cent.

However, the chance of finding a second tumour may depend on how intensively the woman is investigated, and could anyway simply represent earlier detection rather than an increased number of tumours. Furthermore, not all the abnormalities detected in this way would necessarily go on to cause symptoms.

In one interesting study of women under 65 with early (Stage I or Stage II) breast cancers, nearly one in six women had cancerous changes revealed by biopsy of the opposite breast at the time of diagnosis of the first tumour. However, the vast majority of these proved to be pre-invasive *in situ* carcinomas which may never have become symptomatic.

A post-mortem study of women who had died with breast cancer found that over two-thirds had cancerous changes in the other breast, with a roughly equal number of invasive and *in situ* carcinomas.

As discussed in Chapter 3, other studies have found *in situ* carcinomas in the opposite breast in up to 60 per cent of cases of primary breast cancer. Clearly not all of these go on to become invasive cancers, and the significance of these changes remains uncertain, as is the need for treatment.

Whether the opposite breast should be biopsied at the time of initial treatment, or during follow-up, therefore remains hotly disputed. Using random biopsies may detect a high proportion of *in situ* lesions which would never have caused problems if left alone and, conversely, since only a small sample of breast tissue is taken, may well miss some invasive cancers which are present.

To avoid unnecessary anxiety and interference some doctors prefer to monitor the other breast during follow-up visits but to reserve biopsy for cases where some physical or mammographic suspicion exists, and this is largely the case in the UK. However, in the USA there are more advocates for regular biopsy of the opposite breast, some of whom go so far as to argue for prophylactic mastectomy for women at very high risk of a second cancer.

Finally, now that lumpectomy is more often performed in preference to mastectomy, very similar arguments may also apply to the detection of second tumours within the affected breast tissue as well.

Prognosis

There are probably other unknown factors which affect the outcome of breast cancer as well as the ones already discussed. Certainly, even after careful analysis of all the known influences it remains difficult to give a prognosis in any but the most general terms.

For instance, some women with widespread disease who appear to be hopeless cases may defy all predictions and outlive their surgeons. At the other end of the scale, even with small tumours diagnosed at an early stage and treated aggressively with potentially curative surgery and adjuvant therapy, only about 70 per cent of the most favourable cases will actually be cured by the initial treatment.

Overall, it is reasonable to conclude that about one in three women who have treatment for early breast cancer will have a normal expectation of life. However, of the remainder, those who get local recurrences can, as we have seen, often be adequately treated by radiotherapy. Unfortunately only just over half of all the women with breast cancer seen by surgeons have what can be classed as early disease. Nevertheless, most women with breast cancer still have a reasonable expectation of life (unlike the case for some other cancers such as cancer of the lung or stomach) and can expect to live in comfort for many years after treatment even if their disease cannot be classed as 'cured'. Given that breast cancer is predominantly a disease of older women, many of these patients will die from other causes or of old age rather than from their breast cancer.

Breast cancer in men

As has been mentioned before, breast cancer occasionally occurs in men, who form about 1 per cent of all cases. The

Average 5-year survival rates for breast cancer

Overall (all cases)	64%
Stage I	85%
Stage II	71%
Stage III	48%
Stage IV	18%

From CRC factsheet 1988.

smaller amount of breast tissue in men means that lumps are more readily apparent, so the diagnosis is usually made at an earlier stage than in women. However, it also means that spread of the disease to the regional lymph nodes occurs early, so the prognosis tends to be worse in men, who often undergo rapid spread to the rest of the body. In cases caught early, treatment can yield a 70 per cent three-year survival rate, but once there is lymph node involvement this is reduced to under 50 per cent.

The treatment of men with breast cancer follows broadly similar principles to that of women. Some form of definitive surgery, usually mastectomy, is carried out initially. Since men have much smaller breasts and therefore less overlying skin to cover the area after removal of the breast tissue, they may also need a skin graft.

9

Advanced disease

Unfortunately, it is still not uncommon for women to have quite advanced breast cancer by the time they first see a doctor. This may be because they have not noticed, or have perhaps ignored, early symptoms such as a breast lump. Many women are so terrified of the thought of breast cancer that they delay reporting their symptoms. This is a preventable tragedy which should become less common as the importance of early detection is repeatedly stressed in public health education programmes and by the media.

Other women may have an aggressive form of breast cancer which has already spread widely in the body by the time a lump becomes noticeable. Some women have disease which recurs despite the best modern treatment can offer.

When doctors talk about advanced disease they generally mean that the cancer has reached a stage where no treatment is likely to be able to effect a cure. It is important to recognize that although advanced disease is incurable, the natural history of breast cancer is very variable. When a woman has metastatic disease, with widespread secondary tumours, her expectation of life may still range from only a few months to many years. Also, incurable does not mean untreatable. Although the cancer is judged to have spread widely within the body, and nothing can be done to eradicate it, much can still be achieved in terms of reducing symptoms and prolonging the woman's active life for as long as possible.

However, in these circumstances, the more drastic and heroic measures which may be justified for the chance of completely eradicating a cancer are better avoided. They become pointless and merely cause unnecessary discomfort and side-effects. The focus of treatment of advanced disease therefore shifts away from cure *per se*, towards measures which may prolong the length and should enhance the quality of life remaining.

Symptoms of advanced disease

There are broadly two situations in which advanced disease may be suspected, even from the very first time a women sees her doctor with breast cancer. The first is when the cancer is so advanced locally that it is obvious that the outlook is poor and that secondary cancers are almost certainly present elsewhere, even if they cannot be detected. An example would be a woman who is first seen with a large cancer obviously ulcerating through the skin. The second obvious case is when the women already has symptoms of secondary tumours elsewhere, such as bone pain from secondaries in the spine or breathlessness from involvement of the lung. These two examples correspond to Stage 3 and 4 disease respectively (see Chapter 4, pp. 47–8).

Symptoms of locally advanced disease

Locally advanced disease, with an ulcerating lump on the chest wall, is fortunately rare. Even more uncommon is distant skin involvement with ulcers or nodules of tumours beyond the local area of the affected breast, which suggests a particularly poor outlook.

The regional lymph nodes are likely to be affected in advanced disease, and these too can eventually invade and ulcerate the overlying skin.

In severe cases the circulation of lymph fluid draining the arm can be blocked, leading to accumulation of lymph fluid in the tissues (lymphoedema) and swelling of the arm, which may be both painful and inconvenient. Pain and sometimes paralysis of the arm may occur if the network of nerves running through the armpit become invaded by tumour.

Symptoms of spread of disease

When breast cancer spreads from its primary site, certain organs are more likely than others to have cancer cells seed out to create secondary tumours. These include the bones, lungs, liver, brain, and ovaries. Usually the origin of these secondary tumours is apparent, but occasionally women develop symptoms such as jaundice, or spontaneous fractures of the leg bones, before any breast lump is apparent.

Secondary tumours in bones occur often in advanced breast cancer, and it is not uncommon for the first symptoms of the cancer to be either bone pain or a pathological fracture, one which happens without the trauma normally required to break bone. Post-mortem studies suggest that the skeleton may be involved in up to three-quarters of all women who die of the disease, although not all will have had symptoms of bone secondaries. They typically occur in the spine, skull, pelvis, and hip bones. They may be solitary or multiple, and cause either a specific localized pain or generalized aching. Backache is common.

Sometimes a woman experiences weakness and lethargy due either to excess calcium in the blood (hypercalcaemia) because of bone secondaries, or to anaemia (a deficiency of the red pigment haemoglobin in red blood cells) due to involvement of the bone marrow (where the blood cells are produced).

Hypercalcaemia may also cause nausea, vomiting, constipation, excessive thirst, and a general feeling of being unwell.

Spread of cancer cells through the lymphatic vessels into the chest cavity can lead to involvement of the lungs or of the tissue around the heart. A secondary tumour in the lung may cause fluid to leak into the fibrous membranes (pleura) surrounding it, leading to a condition called pleural effusion which is the most common cause of the breathlessness which occurs in many advanced cancers. In a few cases breathlessness may result from blockage of lymphatic drainage around the lungs.

Secondaries in the liver can cause nausea, loss of appetite, and weight loss as well as the intense itching and yellow coloration of the skin typical of jaundice. Often the liver becomes large enough for a doctor to feel it when examining the abdomen. Liver tumours which expand rapidly can produce severe pain.

Secondary tumours in the brain which produce a rise in pressure within the skull can lead to headaches or to specific neurological signs such as disorders of speech or movement and epileptic fits. These may also arise from damage by the tumour itself, which can also cause various disorders of mental or nervous system functions including epilepsy and dementia.

Special tests

The doctor assessing a woman with breast cancer will ask carefully about such symptoms and conduct a thorough clinical examination to detect any signs of secondaries and to make a clinical assessment of the stage of disease.

As discussed in Chapter 4, most women with breast cancer will have a blood test and a chest X-ray, and all should have the tumour confirmed and its grade determined by laboratory analysis, either after some form of biopsy or on a sample removed at operation. For a woman with obvious recurrent disease after treatment however a further biopsy may be unnecessary.

Other specialized tests may be needed to detect secondary tumours or assist in the staging of advanced disease. Involvement of the bones can be assessed by X-rays, and a systematic search of the bones for tumours is known as a skeletal survey. A bone scan, which relies on injection of special radioactive isotopes which concentrate in the bones, may be performed as an alternative.

Ultrasound (high frequency sound waves which reflect back in different degrees according to the type of tissue) can be used to look for involvement of internal organs, especially the liver. Computed tomography (CT) scanning is a specialized X-ray technique producing computer-analysed pictures of cross-sections of the body, and may be performed if there is a suspicion of secondary tumours in the brain.

Treatment of advanced disease

Treatment of advanced disease, whether Stage 3 or Stage 4, no longer aims at cure but at alleviating symptoms (palliation) and at prolonging life where possible. In most cases both objectives can be achieved by the same treatment strategy. However, should the two objectives be in conflict the emphasis will generally be on enhancing the quality of life remaining rather than compromise it by trying to prolong the time left at a cost of distressing side-effects.

Local treatment

For the relatively rare cases of locally advanced (Stage 3) disease, even the most extreme forms of surgery are futile as a life-saving measure. Mastectomy is best avoided even to relieve symptoms since the disease often rapidly recurs within the scar and so becomes even more difficult to control, although in some cases a painful ulcerated lump may be removed to provide temporary relief to a woman with terminal disease. Radiotherapy is generally used in these cases to halt skin ulceration, shrink the size of the tumour, and lessen involvement of the chest wall, although more frequently these days chemotherapy is used for this (see following text). In Stage 4 disease, by contrast, local surgery may be useful to avoid skin ulceration. However, there is clearly no point in performing a hasty mastectomy which may simply add to a woman's suffering if she has a very limited expectation of life.

Systemic treatment

The use of additional non-surgical treatments for advanced disease follows similar principles to those discussed in Chapter 7 for early breast cancer, although the more aggressive regimens are generally best avoided. In about 30 per cent of both pre- and postmenopausal women with advanced disease, adjuvant treatments can halt progression of the disease, alleviate symptoms, or even reduce the size of the tumour. This effect usually lasts for some months, but can persist for years in some cases.

Tamoxifen is generally the first line treatment of choice given to postmenopausal women with advanced breast cancer. Women who have already had tamoxifen and whose tumour has grown despite it may be helped by other hormonal drugs.

For premenopausal women in the UK, the treatment of advanced disease differs from that of early breast cancer, and hormone treatment tends to be favoured over chemotherapy. Traditionally this was achieved through surgical removal of, or radiotherapy to, the ovaries. More recently the drug goserelin (Zoladex) has been given, by monthly injection, to achieve the same effect of halting ovarian oestrogen production, and the use of tamoxifen in these circumstances is now coming to the forefront of treatment of pre- as well as postmenopausal women. A formal assessment of the benefits of tamoxifen in these circumstances should soon be available through the results of a large scale clinical trial now underway.

Tamoxifen may be used for advanced disease irrespective of oestrogen receptor status, although patients with oestrogen receptor-positive status are much more likely to respond. Response rates in women whose tumours are positive for both oestrogen and progesterone receptors may reach more than 75 per cent, compared with only 10 per cent in receptor-negative tumours.

As mentioned above, if the disease recurs or progresses on tamoxifen, it may be possible to go on to use other drugs which affect the body's hormones to produce further benefit. This is known as second line endocrine therapy, and involves the use of drugs called aromatase inhibitors, which antagonize the actions of the adrenal gland (chemical adrenalectomy), so preventing the synthesis of oestrogens. These may be given together with steroid drugs. Progestogens, which offset natural oestrogens, or androgens (the 'male' hormones) are used in some cases, and produce a response in about 20 per cent of women, particular those with secondary tumours in bone.

The advantage of using tamoxifen as the first choice drug for palliation irrespective of a woman's menopausal status is that it is relatively non-toxic and can reduce symptoms of advanced disease considerably. Chemotherapy, by contrast, can be very toxic and may make patients feel awful. In the UK it is generally felt that for most women with advanced disease chemotherapy offers nothing in the way of prolonging life, so is generally best avoided. However, some experienced cancer specialists use chemotherapy to help alleviate symptoms when other drugs have failed, and it is occasionally used in patients with secondary tumours which have become life-threatening, such as those in the liver or lung. Chemotherapy using one drug such as cyclophosphamide can have results as good as those obtained with hormone treatment.

Paradoxically, if chemotherapy is used in such cases, it makes sense to use quite high doses because the increased side-effects are offset by a greater reduction in symptoms of the disease. Thus high dose chemotherapy can yield a better quality of life.

In the USA treatment policies differ and more aggressive chemotherapy regimes are often used in premenopausal women even with advanced disease. While remission rates as high as 80 per cent or 90 per cent have been claimed with cyclical combination chemotherapy, this success is relative and tumour shrinkage is likely to be temporary, lasting perhaps two years at best for each round of successful treatment. Whether the increased chance of a temporary remission is worth the inconvenience and unpleasant side-effects of this type of treatment is a very difficult question. Women whose symptoms remit may enjoy a good quality of life after treatment, but for those for whom chemotherapy fails, the treatment represents an additional burden to that of the disease itself. The options need to be carefully discussed with the woman and her family before such decisions are taken as, unlike hormone treatment, there is as yet no means of predicting which women are likely to respond to chemotherapy.

Bone involvement

Bone pain from a solitary secondary tumour can be resolved in about 95 per cent of cases by a short course of localized radiotherapy. If bone pain is more general then analgesic drugs are needed. Usually mild painkillers such as aspirin or non-steroidal anti-inflammatory drugs like ibuprofen are tried first, progressing up an 'analgesic ladder' of increasing strength to extremely potent drugs such as DF118, morphine, or diamorphine (heroin) if necessary. Where tablets fail to control pain a variety of methods are now available to offer a continuous infusion of drug through a needle under the skin (subcutaneous analgesia) or directly into a vein (intravenous analgesia).

The stronger analgesics often provoke nausea if given alone so an anti-emetic drug is usually given as well to combat sickness. Examples are stemetil or domperidone. Other drugs such as the antidepressant amitriptyline may also be added because these seem to improve pain relief.

Fractures due to weakening of bones by secondary tumours (pathological fractures) require treatment from a specialized orthopaedic surgeon. About half of all secondaries in the femur (thigh bone) lead to fractures and these need to be treated by internal fixation in which the bone fragments are joined together with pins or plates inserted at operation under a general anaesthetic. Some orthopaedic surgeons advocate fixing the femur at the site of a secondary tumour even before it has fractured. A similar fracture may occur in the long bone of the upper arm (humerus), which can be treated non-operatively with a plaster cast or sling but in most cases should be followed by a short course of radiotherapy to the arm.

Sometimes pathological fractures occur within the spine, which has the effect of collapsing the involved vertebra and may therefore put pressure on the spinal cord, potentially leading to dire consequences such as paralysis or incontinence of urine. Such fractures require urgent treatment, either by orthopaedic surgery or by radiotherapy.

Involvement of the bone marrow which produces severe anaemia may require a blood transfusion for short term relief. For longer term control of anaemia symptoms high dose steroids can be given, or hormone treatment, or a particular type of chemotherapy (vinca alkaloid drugs) which attacks the tumour while sparing the marrow cells.

Very rarely, the effect on the bone marrow can reduce the number of blood platelets (thrombocytopenia) which if severe can be life threatening because of widespread bleeding into mucous membranes, and a transfusion of platelets may be required.

Hypercalcaemia due to bone involvement requires urgent treatment to avoid dehydration and impaired kidney function. This involves giving fluids intravenously and stimulating the output of urine (called a forced diuresis) to get rid of some of the calcium. Steroid or other drugs may then be given to stop the problem recurring.

Breathlessness

Shortness of breath from pleural effusion may be controlled in almost all cases by draining the fluid away through a tube inserted between the ribs below the armpit, a relatively minor procedure which can be performed under local anaesthetic.

Chemotherapy drugs can then be introduced through the same tube to destroy the cancer cells on the lung surface and so prevent the effusion coming back. In terminal cases or if the effusion keeps recurring, however, a drain may need to be kept in place almost permanently.

If breathlessness is due to blocked lymph drainage around the lungs this is rather more difficult to treat, but steroid drugs can often offer relief.

Liver involvement

Liver secondaries may be shrunk by high doses of steroid drugs or by radiotherapy to the liver area, and this usually relieves any pain and symptoms of nausea, loss of appetite, or jaundice.

Brain symptoms

Secondary tumours within the brain may also be shrunk by giving steroids, especially a drug called dexamethasone which reduces pressure inside the skull. Radiotherapy to the head may also be effective.

Lymphoedema

Swelling of the arm due to blockage of the lymph ducts is one of the most difficult problems to treat in advanced breast cancer. Some combination of physiotherapy, elastic compression bandaging, and pneumatic compression therapy can reduce symptoms to a tolerable level for most women.

Palliative care

When advanced breast cancer reaches a stage where it is obvious that sooner or later a woman will die, there is no further point in trying to prolong life, and treatment should shift emphasis to the alleviation of symptoms in an attempt to make her as comfortable as possible for whatever time remains to her. Because of the unpredictable nature of the disease, this time may still vary between a few days and a few months.

There is not the scope in a book of this nature for a complete discussion of the principles of palliative care, which have now reached a very sophisticated state. The hospice movement has been at the forefront of the vastly more humane treatment of the dying which has now spread out to hospitals and into the community. The importance of controlling symptoms with whatever dose of drugs is necessary to do so is now recognized. For example, whereas in the past painkilling drugs were often given in inadequate doses or to a fixed time schedule, nowadays it is recognized that there is no need to fear addiction and that powerful drugs such as morphine may be given in adequate doses at intervals timed so that the patient's pain never breaks through.

Pain, insomnia, nausea, constipation, and breathlessness (which is often one of the most distressing of symptoms) and other problems can now be alleviated efficiently with minimal side-effects, while enabling the woman to remain mentally alert for as long as possible. In addition, the aims of management of a dying patient are not just to ensure that troublesome symptoms are kept at bay but also that the patient's mental and spiritual needs are catered for, with her family involved in care where appropriate and their needs and anxieties taken into account as well.

Although the availability of hospice beds and of hospice-run home care services is patchy, it is nowadays possible for a woman and her family to choose hospice or home-based care or to combine the two where desired. This could mean that she spends a few days in a hospice to assess her symptoms and prescribe the correct schedule of drugs to keep them under control. She can then go home to her family under the care of either a domiciliary hospice team or her general practitioner, with further admissions arranged as necessary to stabilize symptoms again or should her family need a respite.

Whatever form of care she chooses, it should now be possible for a woman dying of breast cancer to spend her last days in comfort and to die with dignity.

10

Psychological and family aspects

Many advances have been made in the approach to breast cancer patients in the past decade. The stress and trauma of the whole process of waiting for test results, learning the diagnosis, facing up to treatment, and adjusting afterwards are now treated with much greater sympathy and understanding, which in itself can help to an extent to lessen the shock and sense of loss, anger, and grief understandably felt.

In particular, the technique of frozen section biopsy, whereby women submitted to an operation not knowing whether their lump would prove malignant and therefore they would awake minus their breast, has now been largely abandoned. The move towards surgery which conserves the breast, or at least towards less extensive forms of mastectomy, and the increased use of reconstructive surgery, all help to minimize the physical mutilation which adds insult to the injury of breast cancer.

Nevertheless, psychological reactions are complex, and not all of these advances have yet yielded the expected improvements in the way women adjust to breast cancer. Cancer itself is an extremely emotive topic, and breast cancer has social and sexual consequences affecting a woman's relationships and lifestyle as well as her body image.

The psychological aspects of breast cancer affect all women who might develop the disease, as well as those who already have it, and span a timescale from before it is even detected to the terminal stage when the disease becomes fatal.

Detection of breast cancer

In spite of all the arguments for screening high-risk populations of women for breast cancer, somewhere between 20 per cent and 60 per cent of women offered screening do not accept the invitation. Similarly, despite widespread publicity concerning the importance of early detection of breast lumps and a campaign to promote breast self-awareness, up to 20 per cent of women still present with advanced cancer. Women still go along to doctors with breast lumps which they have known about for a long time but either were too frightened to have assessed or did not think it was anything to worry about.

On the other hand, every breast clinic in the country sees many terrified young women who believe they have cancer when it is quite plain from their age and symptoms that they almost certainly do not. Pain is the most confusing symptom in this regard and many women still believe, contrary to the facts, that breast pain often means cancer or that a non-painful lump cannot be cancer.

Clearly public health education is either not imparting its message or is failing to reach enough women. One important improvement might be to target women at a young and impressionable age, perhaps as part of the school curriculum, to stress that discomfort and even pain in the breasts as part of the monthly cycle is so common as to be regarded as a normal experience, while all lumps detected in the breast should be immediately reported to a doctor.

The current media campaign for breast awareness is an example of strategies which might reach older women. However, much more research is needed to understand the psychological dynamics involved in the fear of cancer, which can lead on the one hand to 'cancer phobia', with unnecessary consultations and needless anxiety, and on the other to the mental block which causes some women to refuse to examine their breasts or to attend for screening *in case* something is found wrong.

Reacting to breast symptoms

When a woman does detect a breast lump or other abnormal symptom, she may broadly speaking react in one of three ways. She may panic, telephone the surgery, and request an immediate appointment that very afternoon. She may realize that this is a problem which needs attention but is probably not, depending on her age and family history, a cause for immediate alarm, and so ask for the earliest free appointment. Or she may be so terrified at the possible consequences that she does nothing at all in the hope that it will all go away.

The curious thing is that the last pattern of behaviour, denial, which is the least appropriate (but very common), does not seem to be influenced greatly by a woman's intelligence or level of knowledge. Some women can delay for years before seeking medical help, yet when questioned as to why, ignorance of the significance of breast lumps barely features as an explanation, nor does fear of mastectomy. Most of these women procrastinate because they are frightened of the possible diagnosis of cancer, not only because of the threat to their life but also because of the disruption it might cause.

Diagnosis of breast cancer

The period of awaiting the diagnosis of breast symptoms is one of intense anxiety for most women, when disturbances of sleep, appetite, and mood are common. Having discovered, say, a lump, which her GP cannot reassure her is normal, and been referred to a surgeon, she is likely to fear breast cancer until proved otherwise. This is when it is crucial for a surgeon to recognize this fear and openly discuss the likelihood of cancer being present, as well as arriving at a definitive diagnosis as quickly as possible. The rapid techniques such as fine needle aspiration cytology described in Chapter 4 are therefore a huge advance from a humane as well as a medical point of view.

If a diagnosis of cancer is confirmed, a woman obviously goes through a period of shock when her emotions may be in turmoil and rapid mood swings, from grief to anger to denial, are common. She is reacting not just to the imminent surgery but to a threat to her life and lifestyle which may seem to extend to her work, her family, and relationships. Although surgery for breast cancer is not urgent in the sense that a few weeks makes little difference, it is kinder to get the period of threat and crisis over as soon as possible by early admission for treatment.

In addition, treatment for breast cancer involves a considerable disruption to a woman's life. She may need to spend a week or two in hospital for surgery and then weeks or months adjusting her schedule around the demands of radiotherapy or chemotherapy treatment. This may cause severe problems for a woman with a young family, and she may need additional help from family, friends, or social services agencies to cope.

After surgery

For many women, coming through breast cancer surgery after all the strain of the previous weeks leads to a curious sense of relief in the first few days. This rapidly dissipates in the early weeks after the operation. Where beforehand was a period of intense anxiety, coping with the diagnosis and preparing to go into hospital, she now has time to think about what has happened and must face the future knowing that she has had cancer.

Women seem to hit a particularly vulnerable period about two or three months after their initial treatment, frequently experiencing a series of reactions very similar to those which follow a bereavement. A period of shock and denial may be followed by anxiety with sleeplessness, loss of appetite, and mood swings as they contemplate what it means to have had cancer. This often gives way to a phase of anger, which may be directed at themselves, as guilt, or at their husband or the surgeon who operated on them. Sadness and depression may follow, particularly if there have been major alterations to a woman's lifestyle or her sexual relationships. Most women eventually arrive at a period of acceptance, and develop a coping style which enables them to get on with life.

However, between 20 per cent and 30 per cent of women still have psychological problems, such as serious depression or anxiety or psychosexual difficulties, one to two years after surgery. This must be compared with the 8 per cent to 10 per cent frequency of such symptoms among women of the same age without breast cancer, but nevertheless represents a considerable psychological burden.

It is difficult to separate the effects of having cancer from the effects of breast surgery in these cases. Until about 15 years ago, when the standard treatment for breast cancer was radical mastectomy, usually with removal of the lymph nodes in the armpit and followed by radiotherapy to the chest wall and armpit, women had to cope not only with loss of their breast but also quite often with a swollen and painful arm due to lymphoedema.

Perhaps not surprisingly, up to 80 per cent of them had problems with psychological adjustment, about one quarter developed anxiety or depression sufficiently severe to warrant medical treatment, and half reported a deterioration in their sex lives.

It was always assumed that these symptoms were largely the result of the loss of a breast. However, surprisingly, as operations which conserve the breast have become more common there has not been the expected huge reduction in psychological sequelae. Anxiety, depression, and sexual problems are common in women after breast cancer surgery irrespective of the type of operation they undergo.

While some studies have shown a reduction in anxiety and depression with lumpectomy, others have shown that after lumpectomy women are often extremely anxious about the possibility of recurrence, even though their psychosexual adjustment may be better. Lumpectomy patients report less of a loss of feelings of attractiveness and femininity, feel less self-conscious about their appearance, and more comfortable talking about their surgery and sexual feelings. They also claim to receive more emotional support from friends, and are likely to rate their partners' sexuality as enhanced after surgery, whereas mastectomy patients more often report a decline.

Yet another study of women taking part in a clinical trial in which the type of surgery they received was assigned at random showed that those who had conservative surgery had just as much anxiety and depression as women who had a mastectomy.

There are a number of possible explanations of this phenomenon. Once a woman returns from hospital after breast cancer surgery, the life-threat of cancer itself can return with a vengeance, and it is possible that anxiety about cancer generally may override feelings about retaining her breast. Conservative surgery may still be too new an approach for women to feel secure that it can 'do the job' in eradicating cancer. It was initially feared that although survival rates were just as good after lumpectomy as mastectomy, the local recurrence rate might be higher, and although with modern adjuvant treatment this has now been disproved, the fear may have stuck in the minds of some women. In addition, the effects of radiotherapy or chemotherapy and the prolonged course of treatment needed after the surgery itself may reinforce the message that a woman has cancer.

However, another interesting observation is that women whose surgeons discuss the treatment choice with them suffer less depression, irrespective of what operation they eventually have, than women not offered the chance of discussion. This supports the idea that the more patients feel involved in their own treatment the better they are likely to adjust.

Sexual relationships

Women, and often their partners, are frequently terrified that a mastectomy will put an end to sexual attraction. Some women, fearing rejection or revulsion, refuse to show their scar to their partner or avoid sexual encounters altogether, which may make this a self-fulfilling prophecy.

Women who are not in a stable relationship are particularly vulnerable because if they avoid social and sexual contacts they may add loneliness and isolation to their existing problems.

Where a woman is married or in a stable relationship it may be an enormous help to both of them to involve her partner at all stages from the diagnosis to the follow-up period. The man may feel guilty about expressing his feelings and may be wary of approaching the woman for fear of upsetting her, and he needs reassurance and encouragement to show his affection and to seek intimacy. What little work has been done in this area suggests that while any major problem can add to the strains of an already fraught relationship, with a happy and stable union the stress of breast cancer and its treatment often actually cements the bonds between a couple.

The follow-up period

Even after the treatment and the period of adjustment to it is over, a woman still has to return to her surgeon at regular intervals for follow-up appointments. No matter how well-adjusted she is, or how optimistic the signs that the disease has been caught in time, each visit is at least a reminder that the disease might recur. In between clinic visits any small twinges or aches and pains which might otherwise be dismissed can reawaken fears of cancer.

If recurrent disease is detected the bitter disappointment can be shattering. Suddenly having to adjust again to the threat of death, especially if this does not happen for some years, can cause severe reactions, and although some local recurrences can be treated and cured, a woman who is found to have secondary spread of the disease must face up to an eventual terminal illness. Studies suggest that up to half of all women in this situation may develop a major depression.

Sources of help

In recognition of the painful catalogue of psychological reactions associated with breast cancer, many treatment centres now employ specialist nurse-counsellors to provide advice, listen to women's fears, and offer support as they go through the process of diagnosis and treatment. They are often present when the woman is told the diagnosis and may be able to visit her later at home to answer questions before she is admitted for surgery. Later, they can tell the woman about other support services on offer and pick up any psychological problems, particularly if they can visit her at home two to three weeks after discharge when she is often at her lowest.

Since they are not mainly focused on the medical aspects of her disease, and can often spend longer talking to her, nurse-counsellors seem better able than doctors to spot psychological reactions serious enough to warrant treatment. However, it is vital that both sides recognize when the time has come to terminate their relationship, as maintaining professional contact on a regular basis can prevent the woman from feeling she is able to cope alone and perpetuate a sense of being 'unwell'. The availability of a friendly voice at the end of the telephone if necessary can, of course, be continued.

Many organizations offer help and advice to cancer patients in general or to women who have had breast cancer (see Appendix). The Breast Care and Mastectomy Association is composed of volunteers who have had breast cancer themselves, who can offer new patients and their families an invaluable source of information, sympathy, and support. Cancerlink offers general advice and literature and acts as an umbrella organization for local self-help groups.

BACUP is an organization founded by a doctor who herself had cancer which offers information and advice by telephone. The Women's National Cancer Control Campaign runs a telephone helpline offering free advice on breast screening, and has a number of free leaflets available by post. The Breast Care Campaign deals mainly with benign breast problems and runs a telephone helpline for women worried about breast problems in general.

Talking about cancer and dying

Many people find it excruciatingly difficult to know what to say to anyone who has cancer or who is dying. Death and dying are no longer a part of everyday life in the way they were to our forebears, when large extended families lived together, death rates were higher, and most people died at home. Cancer and death have become taboo areas, often inciting the feeling that the person affected has somehow been cheated of the perfect health which is often felt to be a modern entitlement, and even that his or her misfortune may somehow be 'catching'. Many people find it difficult to use words like 'cancer' or 'death', and either escape into euphemisms or ignore the subject altogether.

Sometimes people feel that their relative or friend should be shielded from the knowledge that they have cancer, especially if it is clearly terminal. Yet studies have shown that women usually do want to know the truth, and need to know both so that they can make plans and so that they have the chance of coming to terms with it. Sometimes 'shielding' the sufferer is used as an excuse and it is actually the relative who is unable to deal with the issue.

There is a strong argument that women have a *right* to know and, in some countries, this is a legal as well as a moral right. Decisions about future plans may be affected and they may wish to make financial and legal arrangements, as well as perhaps healing any rifts in relationships and deciding how to spend the time remaining. If she is to be asked to consent to treatments with unpleasant side-effects, as is often the case with cancer, she needs to know why such treatment is advised.

Furthermore, women who are not told the diagnosis seem to suffer more depression and anxiety than those who face the worst, and often also feel far more lonely and isolated being kept in the dark. What's more, in many instances they *do* know, but join in the conspiracy not to talk about it for fear of upsetting their relatives. A few people may feel unable to cope with knowing that they have cancer and shy away from asking questions. Certainly, no-one should be forced to confront the truth before they are ready. However, it is crucial that anyone who asks for information should be given it, and generally those who need time to come to terms with their worst fears will make this clear by rejecting details which prove 'too much'.

For people with cancer, one of the saddest experiences can be the feeling that their friends and relatives have become distant and aloof. Both sides can become terrified of mentioning the disease at all, for fear of upsetting the other. The disease becomes an unmentionable subject yet, because it is probably the main thing on everyone's mind, silently dominates all interactions.

The natural resentments which are part of everyday relationships may be stifled because of feelings of protectiveness or gratitude, or from fear of spoiling what remains of life. This too can create an undercurrent of tension and awkwardness.

People with cancer, their relatives, and anyone who has been bereaved, need to talk about their feelings. They need a sympathetic listener who is not embarrassed to hear an outburst of anger, grief, or fear. They need continuing warmth and affection and reassurance. They may need to take things at their own pace, working through initial reactions of shock and denial, but sooner or later most people need to talk.

There are also practical considerations. People going through the shock of having a serious illness diagnosed often forget questions they want to ask, or medical details they are told, and friends and relatives who are not afraid to discuss matters openly can help a great deal by noting things down for them. A person who is facing a terminal illness may need help to make arrangements to put their affairs in order, which is not easy if everyone around them shies away from the subject in embarrassment.

There are no easy formulae to assist communication in these situations, although some tips may help those trying to support someone facing up to breast cancer.

- Let the patient take things at her own pace.
- Respond to prompts that she wants to talk about it.
- Listen without feeling you have to offer advice or solutions.
- Try to talk in a straightforward way without euphemisms.

- Don't be afraid to express your own feelings.
- Don't feel you always have to fill the gaps in conversation, it can help just to *be* there.
- Don't take angry outbursts personally.

11

Risk and protective factors

Breast cancer is not a disease like, say, tuberculosis, where it can be stated with confidence that the tubercle bacillus is *the cause* of the disease, even if other factors make it more likely that an individual will become susceptible to the infection.

Our current knowledge of the causes of breast cancer may be compared with medical understanding of tuberculosis before the responsible bacterium was discovered. We know that some people are more likely to get the disease than others, that it is more common in particular areas, and that it is linked with specific individual characteristics and features of their history and lifestyle.

However, unlike tuberculosis and other infectious diseases, there seems to be no one 'cause' of breast cancer, and the process of its development is not a single step. It appears that some women have an inherited susceptibility to the disease, which then requires the action of one or more environmental risk factors for breast cancer to develop. These genetic and environmental factors interact in largely unknown ways.

The search for the causes of breast cancer is important because it might assist both treatment and prevention of the disease. It could lead to steps to identify women at high risk and offer preventive strategies or screening for early detection. It might enable alteration of any environmental factors found to be partly responsible, or suggestions to change women's lifestyles to reduce their risk.

Clearly, some women are at higher risk of breast cancer than others. However, possession of a set of risk factors does not mean that a woman will definitely develop the disease, nor does their absence guarantee that she will not. It is not clear why some women with particular risk factors develop cancer while others do not, even with the same risk factors.

There are a number of lines of evidence which help in the search for clues about breast cancer development. Researchers can look at how the disease is distributed among defined populations, for example comparing groups or regions with a high or low incidence of breast cancer—this is the science of epidemiology. In the laboratory, they can test samples of breast tissue and breast cancer cells, or look at animal models of the disease, to assess factors which increase likelihood of malignant change, such as chemicals, toxins, or viruses. Clinical research uses a variety of study techniques (see Chapter 1) to compare specific individuals or groups of women with and without disease to try to find statistical correlations with certain characteristics.

In some cases these lines of evidence may overlap, for example where some environmental factor has been shown to increase the likelihood of cancerous change in laboratory samples and is also linked with an increased risk of the disease in humans.

A woman may have more than one risk factor for breast cancer, and broadly speaking the more risk factors she has the higher the likelihood of disease. However, not all factors which have been linked with breast cancer are separate risks. For instance, part of the higher risk of breast cancer seen in developed countries may be because the average age at first pregnancy is higher than elsewhere, while another part may be due to differences in diet.

Doctors often use the terms 'relative risk' or 'risk ratio' to describe the added risk attributable to single factors. A family history of breast cancer, for example, gives a relative risk of 2.0, meaning that someone whose mother, say, had breast cancer is twice as likely to develop the disease herself as a woman with no family history. Absolute risk, on the other hand, denotes the total risk in a set period of time, say, per 1000 women.

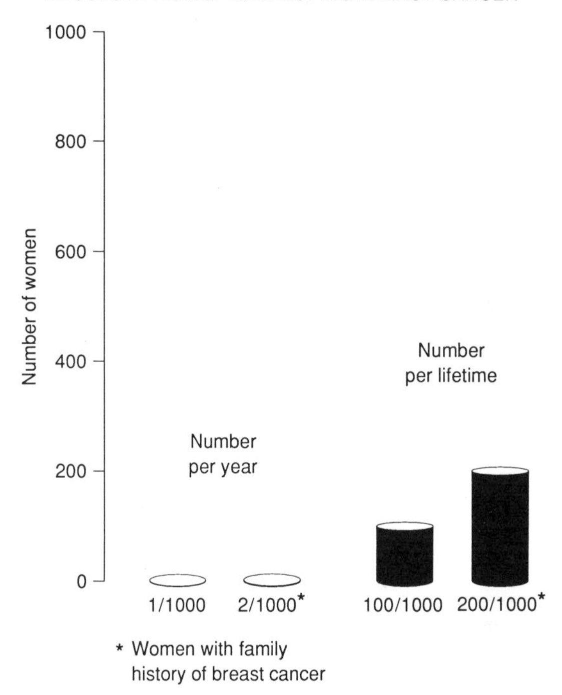

Relative risk of developing breast cancer.

Age

Since certain environmental factors increase the risk of breast cancer, the older a woman gets the greater will be the chance that she will be exposed to these risk factors, and the higher the risk of breast cancer developing. About half of all breast cancers occur in women aged 50–64, and further 30 per cent in women aged 70 or more. Nevertheless the disease does occur in younger women, and it may be that those who do develop the disease earlier in life represent a more susceptible subgroup. Because of the influence of age, any study comparing breast cancer rates needs to assess women of the same age range.

Inheritance

It is often difficult to distinguish genetic from environmental causes. For instance, if two sisters develop the same disease, how do we know whether this is because they share some of the same genes, or because they were brought up in similar circumstances, living in the same places, eating the same foods and so on? Nevertheless a family history of breast cancer is a very strong risk factor, and genetic factors do seem to play a role.

Breast cancer risk factors are probably potentially dangerous mainly to women with an inherited susceptibility to the disease. The risk is much greater if a first degree female relative has had breast cancer, and this is more so if the affected relative is on the maternal side of the family. If a woman's mother, sister, grandmother, or daughter had breast cancer, she has a two-to three-fold increase in her relative risk of developing the disease herself. The risk is particularly great if her relative developed breast cancer under the age of 50, and at the most extreme, a woman whose mother developed bilateral breast cancer before the age of 35 has a nearly 50 per cent chance of developing breast cancer herself.

With the development of new techniques which 'map out'

human genes, scientists are now close to identifying a gene involved in this kind of familial breast cancer. While this represents remarkable progress, it will account for less than 10 per cent of all women who develop breast cancer. For the majority of women without affected relatives it is likely that more than one gene is involved in breast cancer susceptibility.

Also, although a 'genetic screening test' for relatives of breast cancer sufferers might be useful, there are likely to be a number of problems in deciding what to do about a healthy woman with a 'breast cancer gene' which imparts an 80 per cent chance of developing breast cancer in her lifetime.

More aggressive preventive strategies are probably justified in women at very high risk. Some experts have advocated subcutaneous mastectomy with silicone implants as an appropriate prophylactic (preventive) measure. More recently, the role of tamoxifen as a protective agent in such women has been explored.

Geographical variation

There are striking differences in the incidence of breast cancer in different places. Breast cancer is one of the commonest cancers in women in the Western world, but is much less common in Eastern Europe, Asia, and among the black populations south of the Sahara.

Clearly, there are many possible underlying factors which could be involved in geographical variation, including race (linking back to the role of genetic inheritance), climate, diet, environmental toxins, patterns of infection, and cultural influences such as use of birth control, usual age at first pregnancy, and popularity of breast-feeding.

The predominant impression seems to be that the disease is more common among Caucasians living in the colder climates and more highly industrialized countries of the Western hemisphere. This of course does not narrow the field much as far as risk factors go. However, since women from low risk areas such as Japan who move to higher risk countries such as the USA seem to acquire the higher risk of their new country, environmental factors associated with the Westernized lifestyle, rather than ethnic inheritance, seem strongly implicated in determining which women will develop the disease. This argument is supported by evidence suggesting that the incidence of breast cancer increases as countries become more highly developed.

The situation is further complicated though since, in addition to differences in incidence, breast cancer also appears to behave differently in different places.

For instance, among Japanese women who do develop breast cancer, the disease seems to run a more benign course and has an earlier onset than is generally true of Western women. This may suggest some difference in genetic factors regardless of environmental risks.

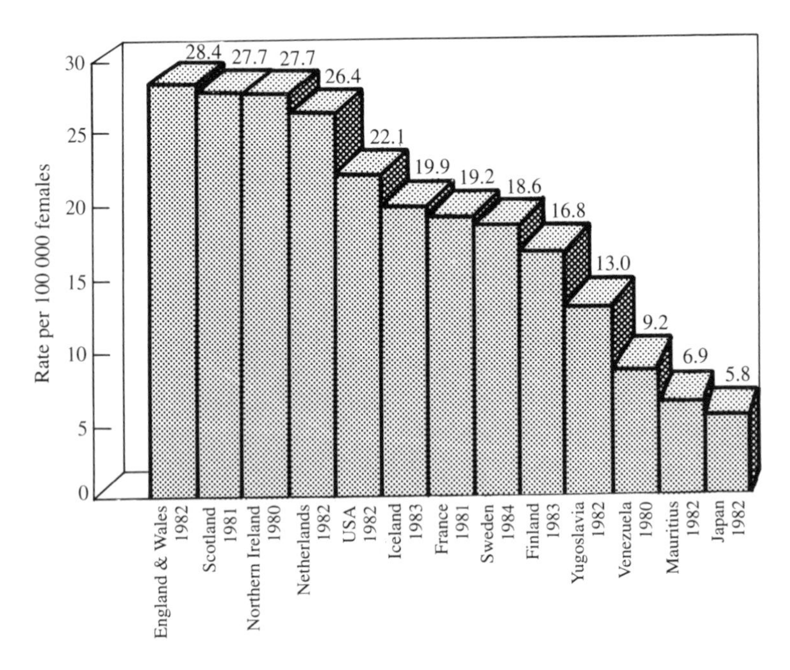

International comparisons of breast cancer mortality.

Furthermore, even within regions there can be considerable variations in breast cancer rates, which may be due to differences in local dietary habits or social class distribution. Women of higher social status seem more prone to breast cancer.

Hormonal influences

Hormones have such a profound influence on breast cancer that they are discussed separately in Chapter 12. Broadly speaking, higher levels or greater activity of oestrogen seem to increase the risk. Pregnancy and/or breast-feeding decrease it, particularly if a woman has her first child at a young age. The greater cumulative total of menstrual periods the higher the risk, so a late onset of menstruation and an early menopause seem to be protective. The oral contraceptive pill and postmenopausal hormone replacement therapy may slightly increase the risk of breast cancer among certain subgroups of vulnerable women.

Other cancers

Women who have had a previous breast cancer, even if apparently 'cured', are at high risk of developing a second primary breast cancer, as discussed in Chapter 8. In addition, women who have had endometrial cancer or ovarian cancer seem to be at higher risk, for reasons which remain unclear.

Diet

The strong influence of factors related to pregnancy and lactation cannot completely account for the large variations in breast cancer incidence between countries. Much attention has therefore been given to the potential role of a 'Westernized' diet in breast cancer risk. Firm evidence for the effects of diet is hard to obtain because it is difficult to dissect out single factors within a nation's or an individual's culinary repertoire, and because people are notoriously resistant to changing their eating habits in the interests of medical curiosity.

Nevertheless, there is no doubt that one of the most noticeable differences between areas of the world where breast cancer is common and those where it is not is in the consumption of animal fats. Comparisons between countries show a strong correlation between reported breast cancer rates and the average consumption of fat and, to a lesser extent, meat. Within countries breast cancer risk is influenced by the degree of obesity.

The evidence incriminating animal fats is sufficiently strong that an extensive experiment is now underway in North America in which a group of women have been asked to modify their intake of animal fat to no more than 15 to 20 per cent of total calories, equivalent to the intake in societies like Japan with a low incidence of breast cancer. These women will be matched with control women of the same age eating a normal 'Westernized' diet containing about 40 per cent of calories as animal fat, and both groups will be monitored for the development of breast cancer. Only if studies like this reveal a difference in breast cancer rates will we be able to say for certain that dietary fat is a risk factor.

People who eat a great deal of animal fat tend to consume rather fewer 'healthy' foods like fruit and vegetables, and a part of the link with breast cancer risk may be due to this rather than to fat *per se*. Recent evidence has suggested that a diet rich in cereals and vegetables, and possibly in dietary fibre, may offer some protection against breast cancer, perhaps through some influence of fibre on the female hormone oestrogen, or through a protective effect of certain vitamin derivatives, particularly vitamin E and beta-carotene.

Obesity

Overweight women have a higher risk of dying from breast cancer than others, and this is not solely because cancers are more difficult to detect in bigger breasts and so are found at a more advanced stage. Apart from the possible influence of dietary factors leading to obesity, in postmenopausal women oestrogen is produced from adrenal hormones within fatty tissue, which offers a theoretical mechanism for this risk.

In addition, there is some evidence that women whose obesity is concentrated around the trunk, as shown by a high ratio between the waist and the hip measurements, are at particular risk. This pattern of obesity, which is linked with a number of diseases, is also more common among infertile women, so it may be that the link is primarily due to fewer of these women having children.

Alcohol

Within the last few years it has been shown that excess alcohol intake, significantly increases a woman's long-term risk of developing breast cancer. The increase in risk is most clear among women who regularly consume more than four units (40 g) of alcohol, equivalent to four glasses of wine or four single measures of spirits, daily.

Both alcohol and oestrogen are metabolized in the liver, and it seems that the liver cells of many women are hypersensitive to the toxic effects of alcohol. After long-term exposure, alcohol damages the capacity of the liver to metabolize oestrogens, resulting in increased levels of biologically active hormone in the blood, itself a risk factor for breast cancer development.

The risk of alcohol needs to be kept in proportion. If one thousand women over the age of 30 each enjoyed a modest amount of alcohol regularly for two years, one extra case of breast cancer might develop. However, this would be offset by reductions in the number of cases of heart disease, and the contribution which enjoyment of alcohol makes to a woman's quality of life.

Most women need not alter their alcohol intake in the light of this finding (although, of course, anyone drinking to excess should cut down for other reasons). However, for women with a number of other risk factors, it may be preferable to confine their drinking to important celebrations and holidays.

Benign breast disease

Women who have had previous breast biopsies seem to have a higher risk of developing breast cancer, mainly where unusual or 'atypical' cells were found in the biopsy sample even if it was judged benign. While some studies have found no links between benign disease itself and breast cancer, others have found the risk to be significantly raised, particularly among women with a family history of breast cancer as well.

As discussed in Chapter 3, some forms of benign breast disease may be extreme versions of changes occurring in the normal breast throughout reproductive life.

In any tissue in which cells regularly grow and regress in a cyclical fashion there may be an increased risk of abnormal growth patterns, and this may increase the possibility of malignant transformation of cells. It is possible that some types of benign breast disease may, in extreme forms, herald premalignant changes in breast tissue.

With the greatly expanded use of breast screening by mammography in recent years, it also seems that some patterns seen on a mammogram may predict the later development of cancer, even though they are not cancerous at the time of the screening. However, these minimal changes require specialist interpretation and remain controversial. There has as yet been insufficient follow-up of women with abnormal, but not cancerous, mammogram patterns for doctors to know what is the chance of cancer developing or of the condition resolving.

Viruses

In certain strains of mice, breast cancer is extremely common and can be transmitted from mothers to daughters. Female offspring almost invariably develop the disease, unless they are separated from their mothers and fed artificially or by a mouse 'foster mother', when they generally do not develop breast cancer. Investigations have shown that milk from these strains of mice contains particles of a virus known to produce an enzyme (a substance which assists chemical reactions in the body) which can create cancerous changes in body cells.

There is little evidence that human breast cancer is associated with viruses or is transmissible through breast milk from mother to daughter.

However, there is an isolated community of Iranian women who have a one in five chance of developing breast cancer in their lifetime. One study of these women has demonstrated particles similar to the mouse breast carcinoma virus in their milk. No study has yet been published assessing whether breast cancer rates vary between women who were breast- or bottle-fed as babies. Such a study could throw interesting light on any possible associations between breast cancer and early exposure to both viruses and environmental toxins.

Radiation

There is little doubt that ionizing radiation may promote the development of breast cancer. Women who received repeated high dose chest X-rays while being treated for tuberculosis in the past have been shown to be at increased risk. Japanese women exposed to the radiation of the atomic bomb attacks on Hiroshima and Nagasaki still develop breast cancer at much higher rates than expected compared with Japanese women of the same age living elsewhere. Furthermore their risk of developing cancer is directly related to their radiation exposure as measured by the distance they were from the epicentre of the blasts.

The effect of radiation in promoting cancers is dose dependent and cumulative, which is to say that the higher the dose and the higher the number of exposures the greater the risk. There seems to be no safe threshold of radiation at which we can say the risk disappears. This is why all radiation is handled with care and should be treated with respect.

As far as the uses of radiation in medicine are concerned,

the risks of, say, a single ordinary X-ray are so small as to be considered negligible.

Nevertheless, repeated exposure to X-rays could produce accumulated doses, and this is one reason why there is a slight anxiety about intensive breast screening by mammography, involving repeated X-rays of the breast. However, this potential risk should not be exaggerated compared with the possible benefits of a screening programme.

Type of breast tissue

There may be factors within an individual woman's breast which have a bearing on the risk of cancer. In recent years researchers have suggested that some of the non-glandular supporting cells within the breast may influence the growth of cancer cells. These cells, called stromal fibroblasts, form part of the supporting structure of the breast and produce a number of chemical messenger substances called growth factors which seem to 'communicate' chemically with breast cancer cells to influence both their growth and their ability to spread. Some women's fibroblasts may be more likely to support the growth of cancer cells, and this may partly explain the hereditary factor in breast cancer. It has also been shown that similar cells in the skin of breast cancer patients with a strong family history of breast cancer behave in a particular way, as do skin fibroblasts of their near relatives unafflicted by breast cancer. This could form the basis of a test for breast cancer risk, although such tests are still very much at a research stage.

Other growth factors suppress breast cancer cell growth and their production is stimulated by tamoxifen.

Personality

There is little evidence to support the commonly held belief of a cancer-prone personality. However, one study has supported the popular belief that bereavement and other loss experiences, such as illness or redundancy, may make people more prone to cancer or more likely to get recurrences after treatment. The postulated reason for this is that stressful experiences depress the immune system, giving cancerous cells more chance to take hold.

Since many more studies have failed to find such links, such suggestions should be interpreted with caution. One study even found that women without breast cancer experienced more stressful life events than women with the disease. On current evidence any links between personality or stressful life events and breast cancer development remain, at best, unproven.

Other links

Many other factors have been suggested as potentially linked with breast cancer risk, although their importance is hard to quantify. They include deficiency of thyroid hormone (hypothyroidism), immune deficiency, tuberculosis, and possession of a particular type of ear wax (wet, as against dry, cerumen). Interestingly, women with AIDS do not have a higher breast cancer risk, which seems to refute the popular belief that depression of the immune system contributes to the disease.

Protective influences

Many of the known breast cancer risk factors are not amenable to alteration. There is nothing a woman can do to alter her family history (unless genetic engineering makes major leaps forward) or the age at which she starts to menstruate. Few women would wish to have children at an early age solely to protect against breast cancer.

Breast-feeding is one easily available form of protection once women do have children, and this is a valid additional reason, on top of all the other benefits for mother and baby, to urge as many new mothers as possible to breast-feed.

Since the development of a new breast cancer is a relatively rare event, affecting only one or two women per thousand per year, any preventive measures targeted towards women at large must affect many for the possible benefit of only a few. For even the simplest and least harmful of manipulations, this introduces a complicated balance of benefits and hazards which need to be carefully thought through before any attempt is made to alter the lifestyle of, or offer drugs to, women in general.

Smoking

Women who smoke seem to have lower rates of both breast cancer and benign breast disease than non-smokers. While for obvious reasons smoking cannot be advocated as a preventive measure, this observation may shed some further light on the process of development of breast cancer.

The protective effect has been repeatedly observed and seems to apply only to current, not former smokers. There are a number of suggested explanations, of which perhaps the strongest candidate is that smoking exerts an anti-oestrogenic effect, which also explains its links with some other diseases. It also accelerates the onset of the menopause, and the protective effect seems to be stronger among postmenopausal than premenopausal women. Smokers tend to be thinner than non-smokers, and some oestrogens are made within fatty tissue.

The reduction in benign breast disease among smokers, which is independent of age at menopause, could partly explain a reduced breast cancer risk, particularly since it is most noticeable for the type of fibrocystic changes associated with premalignant disease, and for duct papillomas, which also have the potential for malignant transformation. However, smoking does affect the tiny arteries in the breast and may lead to recurrent inflammation.

The anti-oestrogenic effect of smoking could be due to decreased production, decreased activity, or increased breakdown of oestrogen. There is some evidence that it may have similar effects to drugs called aromatase inhibitors used in treatment of advanced disease to prevent oestrogen synthesis from hormones released by the adrenal gland. One study found that smoking women have higher levels of androgens, hormones like the 'male' hormone testosterone which are normally present at low levels in women.

If the reason for this link with tobacco could be found it might give further clues to possible prevention with less harmful agents. Meanwhile, however it is clear that the health disadvantages of smoking, not least the risk of lung cancer, far outweigh the small protection against breast disease.

Diet

Despite its disadvantages, a Western lifestyle probably has more advantages than disadvantages, both overall and in health terms. The most strenuous manipulation of her diet to cut down on fats, increase fibre, and avoid alcohol is likely to yield only a minimal reduction in a woman's breast cancer risk. Persuading large segments of the population to alter their diet and lifestyle is notoriously difficult and even if successful the effects may not be as anticipated.

Nevertheless, a woman at high risk of breast cancer would be sensible to avoid obesity and to keep her intake of animal fats and alcohol at a reasonable level.

Tamoxifen

The drug tamoxifen, used in the treatment of breast cancer, has been proposed as a means of reducing the risk of breast cancer among women at high risk, particularly those with a strong family history of breast cancer striking young.

Trials of tamoxifen as adjuvant therapy have demonstrated that not only does it reduce the risk of relapse and death, but also that after long-term administration it reduces the risk of getting a second cancer in the other breast by nearly half.

The problem with using tamoxifen more generally in the prevention of breast cancer is that although its reported side-effects are few when used in treatment, use as a preventive measure could imply much longer-term treatment. This could incur unanticipated long-term effects, and obviously this becomes a much larger consideration if it is proposed to give a drug to large numbers of healthy women. What is acceptable for women with a disease, or even for women at high risk, may not be so for the general population, unless it has some other benefit.

What makes the prospect of preventive treatment with tamoxifen so exciting is that it may well have more general benefits. A recent study in Scotland looking at the use of tamoxifen in early breast cancer noted that the drug also led to a significant reduction in deaths from heart disease.

This has been explained by a number of studies which have shown that tamoxifen significantly reduces cholesterol levels in the blood, and in particular lowers the levels of a type of cholesterol-carrying protein called low density lipoprotein (LDL) thought to be closely associated with the development of arteriosclerosis, the narrowing of the arteries which can compromise blood supply to the heart muscle and lead to heart attacks.

Other researchers have been assessing the effects of tamoxifen on the bones of older women. After the menopause the lowering of body oestrogen levels can lead to a loss of bone mass and ultimately to bone thinning, a disease called osteoporosis. Studies in London and the USA have shown that tamoxifen seems to protect the skeleton from osteoporosis, which could have major benefits in reducing the number of fractures in elderly women, one of the most important causes of loss of mobility in this group.

These findings are so promising that the UK Cancer Research Campaign and Imperial Cancer Research Fund have now jointly set up a trial which will recruit 15 000 women judged to be at increased risk of breast cancer. They will take either tamoxifen or an inert placebo pill for five years, and the incidence of breast cancer in the two groups will be compared. Similar trials are underway in the USA, Italy, and Australia and the results of these studies should give firm evidence about the costs and benefits of tamoxifen prevention, at least among higher risk women.

Other drugs

Despite the current fears that the contraceptive Pill may be linked with a tiny increase in risk of breast cancer, the possibility of using Pill-type hormones to reduce breast cancer risk by reducing secretion of natural oestrogens is theoretically plausible. No such agent is yet available, but attempts to produce such a drug are the focus of intensive research.

Secondary prevention

The use of measures which try to alter the known risk factors for the development of a disease is known as 'primary prevention'. An alternative approach is to try to prevent the effects of the disease rather than its development. For breast cancer, this means breast screening to detect and treat cancers as early as possible, so reducing the risks of spreading disease and death. Early detection and the use of screening mammography are discussed in Chapter 13.

Conclusion

What we know about breast cancer suggests that it develops when a woman who has an inherited susceptibility to the disease is exposed to one or more of a variety of promoting factors which may, at some point during the course of her lifetime, initiate a genetic change leading to malignant transformation in the cells of her breast. There may be a final common pathway through which these risk factors operate in susceptible women, probably associated with high levels of free, biologically active oestradiol, the main circulating form of oestrogen. Means to identify women at high risk and to manipulate these hormones may offer the best means of preventing the disease, and could provide additional health benefits.

12

Hormones and breast cancer

As has been discussed in other chapters, the hormones within a woman's body have a profound effect on her risk of getting breast cancer, on the course of the disease should she develop it, and on the outcome of some forms of treatment. Both the characteristics of her hormones and their fluctuations in the normal physiological states of menstruation, pregnancy, and lactation are important. It also becomes obvious to ask whether the contraceptive Pill or hormone replacement therapy (HRT) could influence breast cancer risk.

The hormonal environment

Breast cancer has been described as a hormone-dependent tumour, and it has been shown that women with breast cancer have more biologically active female hormones than others. The varying proportions of the different types of oestrogen circulating in the blood stream may affect breast cancer risk, and the cells of some tumours have on their surface specific oestrogen receptors whose presence influences the likely success of hormone treatment. In addition, although breast cancer in men is rare, among those who do develop it there seems to be a link with abnormal oestrogen levels.

Women who have an artificial menopause (through surgical removal of the ovaries or drug treatment to stop the ovaries producing oestrogen) before the age of 35 seem to have a lower risk of breast cancer. As a woman passes through a natural menopause, when levels of oestrogen reduce dramatically, the annual risk of breast cancer dips, before increasing again later on.

Although the risk of breast cancer increases with age, the rate of increase seems to be slower among postmenopausal women, who have much lower oestrogen levels than premenopausal women.

In a fascinating study on the island of Guernsey, urine samples were collected from all the women and stored for later analysis. Whenever a woman developed breast cancer, her urine was analysed and the results compared with those from a control group of urine samples. Abnormal hormone metabolites were found in urine up to ten years before the development of breast cancer.

Pregnancy and lactation

Breast-feeding has long been believed to have a protective effect against breast cancer, although recently it has been shown that pregnancy is a more important influence. These two factors are obviously difficult to separate, but there is some evidence of an independent effect of lactation. Firstly, dairy cows virtually never develop cancer of the udders. Secondly, the boat women of Hong Kong have the unusual habit of suckling their babies almost entirely from the left breast, in order to keep their right hand free for rowing their boats. If they do develop breast cancer it is more often in the right breast which has not been used for lactation. A recent study in the UK has confirmed that breast-feeding, even for a short time, is protective.

Having children protects women against breast cancer, and the younger they have them the better from this point of view. Recent evidence suggests early pregnancy is protective whether women breast- or bottle-feed their infants. Women who have their first pregnancy later in life, and those who remain childless are at a significantly increased risk, which may partly explain why infertility is linked with breast cancer among older women.

It may also be one reason why breast cancer rates seem to be increasing in developed countries, where more and more women are delaying having children until they are in their thirties. Women who have their first child under the age of 25 seem to have about half the risk of breast cancer of those who have their first child over the age of 30, or those who never have children. Probably for this reason, nuns seem to have a high incidence of breast cancer.

It seems to be the first pregnancy which confers the major protection against breast cancer, and to do so it must be carried to term. Some studies have shown small further risk reductions with an increasing number of children, but others have shown no effect of subsequent pregnancies after the first. A first pregnancy which ends in abortion or miscarriage seems to have no effect.

Menstruation

It may not even be pregnancy *per se* which is protective. Both early age of onset of menstruation (menarche) and later age at the menopause seem to increase the risk of breast cancer. This could contribute to the increasing incidence of the disease, as the average age of starting menstruation is reducing and the average age of the menopause increasing in Western countries. It may be that it is the cumulative total of menstrual cycles which matters.

It is only relatively recently in human history that women did not spend the greater proportion of their reproductive years either pregnant or lactating, and that they lived long enough to reach the menopause at all. It has been suggested therefore that in biological terms prolonged menstruation may be an abnormal state, and that this may therefore predispose to breast cancer.

The number of menstrual cycles before the first pregnancy may be the ultimate determinant of breast cancer risk. This may further explain the difference in incidence between Western and developing countries. An average well-nourished English girl now starts menstruating at age 12, but the average age of first pregnancy is now 25.6 years, so she will menstruate for more than 13 years before becoming pregnant. An undernourished Asian girl may not begin menstruating until 17 or 18, and could become pregnant almost at once.

Breast cancer in pregnancy

Some 15 per cent of all breast cancers arise in women of childbearing age, and about 3 per cent of patients diagnosed as having breast cancer are pregnant or breast-feeding at the time. It is a rare event (between 10 and 39 per 100 000 pregnancies), but the proportion is likely to increase as more women delay having children until in their thirties and forties, when they are also at a higher risk of breast cancer.

Breast cancer in pregnancy is obviously a particularly tragic combination of circumstances. In the past doctors often had a very negative attitude, believing that breast cancer was more likely to arise during pregnancy, that the hormonal changes of pregnancy and the youth of the patients made the disease more aggressive, and that the prognosis was far worse than for non-pregnant women.

It now seems that breast cancer is no more likely to arise and behaves no differently during pregnancy than at any other time. The hormonal changes of pregnancy do not seem to make the situation worse, there is no difference in the proportions of different types of tumour and, as with other premenopausal women, most tumours are oestrogen receptor-negative.

However, the disease is likely to present at a more advanced stage than the average, and it is probably this which is responsible for the admittedly poorer overall outlook of tumours diagnosed in pregnancy. Various studies have shown that between 56 per cent and 90 per cent of pregnant women with breast cancer have involved lymph nodes. When compared with non-pregnant women of the same age and stage of disease there is no difference in five-year survival rates. This implies that rather than being more aggressive, breast cancer in pregnancy is simply detected later than normal.

There are several possible reasons for this. The normal

enlargement and feeling of fullness of the breast in pregnancy may tend to mask any lump. The feel of the breast has changed and women may not attach the same significance to any lumps which do appear. If a lump is detected, both women and their doctors may tend to wait and see what happens until the baby is born or the woman stops breast-feeding, as many breast changes resolve then anyway. Mammography is less likely to be performed during pregnancy because of the risk of X-rays to the fetus, and even if it is the denser breast tissue of pregnancy makes it less likely that abnormalities will be detected.

Whereas termination of pregnancy used to be recommended routinely if a woman developed breast cancer in pregnancy, this is no longer the case. Indeed, one study has suggested that abortion actually reduces the mother's chance of survival. However, if radiotherapy or chemotherapy are recommended during the first three months of pregnancy the doctor may suggest an abortion to avoid the risks of damage to the fetus.

Also some women may wish to consider this option either because they cannot cope with both pregnancy and cancer treatment, or because they have particularly aggressive or advanced tumours with a poor outlook. Clearly the risks of leaving a young child without a mother need to be carefully considered by the patient and her partner and family.

Breast cancer in pregnant women is generally treated in exactly the same way as in any other women, with initial surgery perhaps followed by adjuvant treatment, with the exception that radiotherapy should be delayed until after pregnancy, or a mastectomy rather than wide local excision performed to avoid the need for radiotherapy.

The risks to an unborn fetus whose mother undergoes radiotherapy are probably quite high, particularly in the early stages of pregnancy. However, much of our knowledge about the effects of radiation come from studies of Japanese women exposed to atomic bomb blasts while pregnant, and the precise risks of therapeutic radiation in pregnancy have not, for obvious reasons, been quantified. In some cases doctors may recommend an abortion for an early pregnancy where radiotherapy is desirable.

We know much more about the effects of chemotherapy, although the numbers studied remain quite small. It seems best to avoid toxic drugs in the first three months of pregnancy. Thereafter it seems that selected drugs may be given with safety although, as with radiotherapy, whether there are long-term effects on the offspring (for example, when they become adults or have children themselves) is unknown.

The use of hormone treatment as adjuvant therapy in breast cancer is becoming increasingly widespread, even in premenopausal women. Little is known about the effects of tamoxifen or goserelin in pregnancy, however.

There is no risk to a fetus from a pregnancy occurring after treatment for breast cancer has finished, whether this has involved radiotherapy or chemotherapy. However, a woman's fertility may have been compromised by past treatment. If a woman who has had breast cancer does become pregnant (and about 7 per cent of women who remain fertile after treatment do so), this does not seem to increase the risk of recurrence of the tumour. Indeed, there is some evidence that such women have a slightly better outlook. However, most doctors would advise women to wait for two or three years after their initial treatment, until they are over the period when most recurrences occur.

The contraceptive Pill

There has been intense public concern generated by several well-publicized suggestions that use of the contraceptive Pill might increase a woman's risk of breast cancer. The issue is extremely important because the Pill is the most popular reversible form of contraception in many countries. Since its introduction over 30 years ago, the Pill has been used by about 150 million women. So even a small increase in breast cancer risk attributable to the Pill could affect large numbers of women.

Despite many large-scale studies involving tens of thousands of women, the issue of whether or not the Pill is linked to breast cancer has been one of intense controversy. This is perhaps in itself reassuring, for any substantial risk would have been readily apparent.

Nevertheless there remains concern about a small increase in risk after long-term use, particularly among certain subgroups of vulnerable women.

This applies only to so-called combined Pills, which contain both oestrogen and progesterone. The progesterone-only type of Pill (which is slightly less effective but avoids some of the side-effects of the combined Pill which are attributed to its oestrogen content) is not implicated in breast cancer.

The combined Pill has a variety of effects on breast tissue which need to be considered. Its action as a contraceptive is brought about by giving synthetic oestrogen which effectively fools the body into believing that the oestrogen has come from the ovary. The normal feedback mechanisms controlling ovulation are therefore disrupted and the ovaries cease their normal monthly egg release. The effect has been likened to that of pregnancy which, as discussed above, may be a more 'natural' biological state than regular menstruation.

However, although production of natural ovarian hormones is decreased, this is compensated for by the added effect of the synthetic hormones in the Pill, and it is the synthetic oestrogen in the combined Pill which has caused most concern in terms of its effects on the breast.

The breast tissue of women on the Pill undergoes cyclical changes very similar to those of women not on the Pill. However, on the Pill the monthly growth and proliferation of breast tissue both occurs at a higher rate and lasts for longer, extending into the first half of the cycle (days 6 to 13) as well as the second, premenstrual phase.

When women are taking the Pill the monthly bleed is not in fact a 'proper' menstruation at all, since it does not follow the release of an egg and failure of the egg to be fertilized. Rather it is what doctors refer to as a withdrawal bleed, meaning that it happens in response to changing hormone levels in the week when the Pill is not taken.

The increased proliferation of breast tissue is what causes concern about breast cancer as it is theoretically possible that it could make the changes which lead to cancer more likely. However, it is unlikely that the Pill could initiate the changes which lead to breast cancer; more likely it could act as a promoter, perhaps of pre-existing disease or premalignant changes.

The many studies which have compared breast cancer rates among women who have taken the Pill and those who have not have given conflicting results. Some early studies which suggested an elevated risk have been criticized because older forms of the combined Pill contained a much higher oestrogen dose, so their results may not be applicable to modern preparations. Furthermore, several large studies have failed to demonstrate any association between taking the Pill and subsequent development of breast cancer, and some have shown a reduced incidence of benign breast disease among Pill-users, which might be expected to reduce the risk of breast cancer.

Nevertheless, some studies have revealed a slightly increased risk attributable to the Pill, particularly among women with a family history of breast cancer or those who have had benign breast disease. A recent large study from Oxford has suggested that women who start taking the pill when very young and continue to take it for a long time (over 8 years) before their first pregnancy do have an increased risk.

However, this must be kept in perspective. The risk amounted to something like a three in 1000 chance of developing breast cancer under the age of 35 while taking the Pill, compared with two in 1000 for women who have never used the Pill. This is rather less than the risk of being run over while crossing the road, although it remains unknown whether the risk persists beyond the age of 35, and since most breast cancers occur in older women this may be more important.

Also, the links between the Pill and breast cancer cannot be viewed in isolation. The Pill seems to have effects, both positive and negative, on a variety of other conditions. For each woman the risks and benefits must be set against her own personal and family history and risk factors for these conditions. For example, for a woman with a high risk of ovarian cancer, the protective effect of the Pill against this disorder may outweigh the small risk of breast cancer. Conversely, for women who have had a clot in the leg veins (deep vein thrombosis), taking the Pill substantially increases the risk of another one and should probably be avoided.

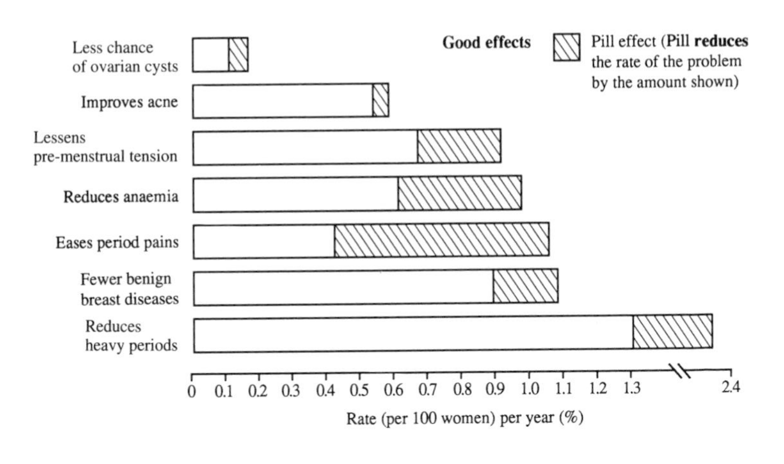

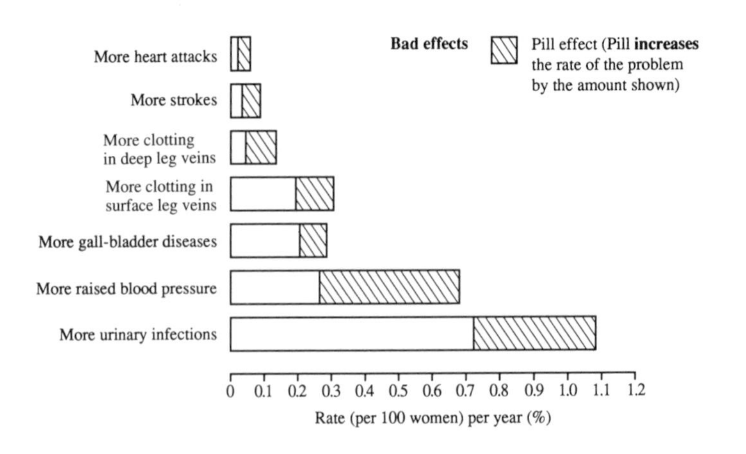

Links between the pill and various conditions.

Should breast cancer develop while a woman is taking the Pill (as it may sometimes, simply because breast cancer is not uncommon and many women take the Pill, so some would succumb anyway irrespective of the Pill), she will be advised to stop taking it. The cancer can then be treated in exactly the same way as it would if she had never taken the Pill, including the use of hormone treatment if needed. It is unlikely that use of the Pill could alter the growth rate or behaviour of a tumour, but most doctors would probably advise other methods of contraception after treatment for breast cancer.

Finally, it must be remembered that the effects of the Pill have been subject to so much scrutiny that probably more is known about its safety than about that of any other medication. In addition to the medical pros and cons, the contraceptive protection offered by the combined Pill (about 99 per cent, better than any other form except sterilization) makes a significant contribution to a woman's quality of life as well as reducing unwanted pregnancies and possible abortions from the failure of other methods. On balance, the benefits of using the oral contraceptive Pill for a sexually active woman who wants effective and convenient contraception probably outweigh the risks.

Hormone replacement therapy

Around the time of the menopause, a woman's ovaries stop producing oestrogen. The variation in hormonal levels leads to irregular menstruation and may cause hot flushes and vaginal dryness. Other symptoms commonly attributed to the menopause, such as depression, irritability, headaches, increased frequency of passing urine, and skin problems are actually no more common in menopausal women than others, and are probably not due to hormone imbalances but to increased awareness because the woman feels 'menopausal'.

Lack of oestrogen in postmenopausal women affects the efficiency with which bones renew their structural elements and this is the main reason why women gradually lose bone mass after the menopause. Eventually this bone thinning results in a condition called osteoporosis, in which bones are so brittle that they fracture easily after even minor falls.

Hormone replacement therapy (HRT) is increasingly prescribed to women around the time of the menopause both to lessen menopausal symptoms and to protect against osteoporosis. When it was first used more than 50 years ago only oestrogen was given, but modern HRT preparations usually consist of a combination of oestrogen and progesterone. This is because when oestrogen was given alone it was linked with a rise in cancers of the lining of the womb (endometrial cancer), but this is not a risk if it is given with progesterone.

We know from studies with animals that giving oestrogen increases the risk of breast cancer. As discussed above, breast cancer risk is higher in women who start to menstruate young and whose menopause is later, in other words in those women who have the longest exposure to natural oestrogens. Consequently the fear that HRT could increase breast cancer risk is a reasonable suspicion.

However, there does not seem to have been a dramatic increase in breast cancer since the use of HRT became more widespread some 20 years ago, suggesting either that any risk must be quite small or that if HRT does promote breast cancer it takes a long time to develop.

Many studies have now looked at the risk of breast cancer among women who have used HRT. As with the Pill, none have shown alarming rises and a number have failed to show any increased risk at all.

However, overall there does seem to be a slightly elevated risk (a relative risk of about 1.06, a 6 per cent increase), and again this appears to be concentrated in certain subgroups of women—those with benign breast disease, those with a positive family history, and those who have used HRT for a long time, more than ten years. In these subgroups the relative risk of breast cancer seems to be about 1.3, that is a 30 per cent increase.

However, the issue is somewhat complicated. For one thing, women on HRT are more likely to be of a higher social status, and therefore are at greater risk of breast cancer anyway. The detection rate for breast cancer may be increased among women on HRT, either because of increased awareness prompting them to examine their own breasts, or because they more often come into contact with doctors who may be more likely to examine their breasts. In addition, mammographic screening is recommended at the start of HRT in women over 50, and every three years thereafter. This is the same as the recommended screening programme for women at large, but not all women accept invitations to screening whereas women on HRT are much more likely to be screened if it is a prerequisite for treatment.

Results of four studies which assessed breast cancer mortality suggested overall that breast cancer risk was raised about 1.4 times after taking HRT for eight years. However, all four studies also showed an improved survival from breast cancer in these women. This suggests that the increased frequency of breast cancer could be explained by surveillance bias, in that women taking HRT are more likely to have their cancers detected.

This would imply that such cancers are more often diagnosed at an earlier stage when they are more amenable to treatment. Women on HRT who do develop breast cancer have been shown to be more likely than other breast cancer patients not to have involved lymph nodes, and to be oestrogen and progesterone receptor-positive at the time of diagnosis, making a response to hormone treatment more likely.

Another complication is that in assessing the long-term effects of HRT, studies must necessarily include women who were given HRT containing only oestrogen, and using higher oestrogen doses than are used in today's combined oestrogen-progesterone preparations. This could also bias the results and perhaps make it more likely that studies will show increases in risk which may not apply with the use of modern preparations.

Furthermore, just like the Pill, any risks of HRT must be balanced against definite health benefits, both in terms of relief of menopausal symptoms and protection against other diseases. HRT not only protects women against osteoporosis, but also reduces the risk of heart disease, to which they become more vulnerable after the menopause. Overall, women who have used HRT seem to have a lower average mortality rate at any given age than those who have not.

Generally HRT is not recommended for women who have breast cancer, as we do not know whether it could increase the risk of recurrence. However, if a woman has severe menopausal symptoms she may be given HRT, provided it is in a form in which oestrogen is combined with progesterone and that she is under the care of a breast cancer specialist.

In summary, the use of HRT in postmenopausal women appears to be relatively safe, although its use in women at high risk of developing breast cancer should be carefully monitored. It may be prudent to limit the length of time HRT is taken, particularly in women at high risk of breast cancer, say to 10 or 15 years rather than continuing treatment for life.

13

Screening and breast awareness

Detection of early breast cancers

Breast cancers which are small and have not spread at the time they are first diagnosed are more likely to be curable. Therefore any means of detecting tumours while they are still at this stage could, in theory, improve the outcome and increase the chance of survival.

One possible way to increase early detection is to encourage women themselves to be alert to changes in their breasts and to report immediately to a doctor any symptom which might herald breast cancer. Another possibility is screening, in which large populations of healthy women are examined or tested medically to try to detect changes even before the woman has any symptoms.

Mammography

The screening programme for breast cancer now in use in the UK is based on mammography, low dose X-rays of the breast which reveal areas of abnormality within the breast tissue. Changes on a mammogram are not necessarily cancer, as a variety of benign conditions show up as well.

Typically, on a mammogram benign lumps appear uniform, rounded, and with smooth edges, and may be surrounded by a 'halo' which represents an encircling layer of fat. Cancerous lumps are more dense in the centre than at the edges, which are irregular, and are accompanied by evidence of other changes such as skin thickening and distortion of the breast tissue around the lump.

Particles of calcium are deposited in the area of many long-standing breast abnormalities, a process called calcification which occurs in about 30 per cent of both benign and malignant tumours. However, whereas the calcification of a benign lesion appears as relatively large, coarse blobs, with cancer it assumes a fine, speckled appearance which to the experienced eye is virtually diagnostic of malignancy.

The mammogram itself involves squeezing each of the breasts in turn between two metallic plates to take the X-ray pictures. Some women find the pressure on the breast uncomfortable, but the procedure only takes a few minutes. The X-ray picture can be processed very rapidly and the woman is usually told the results within four or five days, when she will be asked to return for further tests if any abnormalities (whether benign or malignant) have been noted.

Because of associated changes in the breast tissue which occur

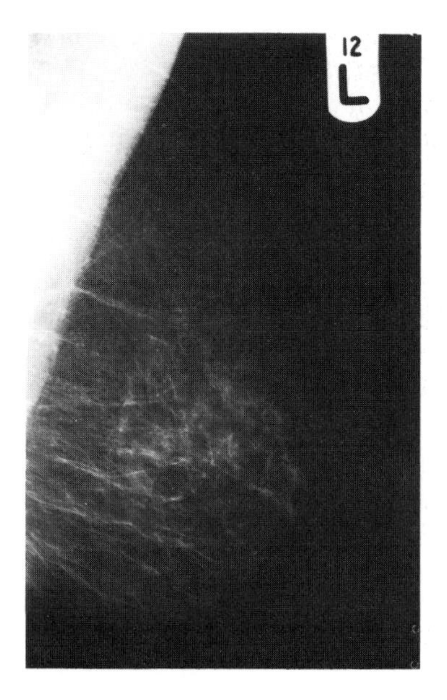

Normal mammogram.

with tumours, mammography can enable detection of cancers less than 0.5 cm in diameter, whereas a lump cannot be detected by clinical examination until it has grown to over 1 cm across, and the average lump detected by feel alone is already 2 cm in diameter.

It might therefore seem obvious that all women should regularly be screened by mammography to pick up cancers as soon as possible after they form. Unfortunately the situation is not quite that simple, since there are side-effects, costs, and disadvantages of the procedure as well.

Evidence for the use of mammography as a screening procedure

The effectiveness of screening by mammography has been under assessment for more than two decades, for a large part of which it has been the subject of some controversy.

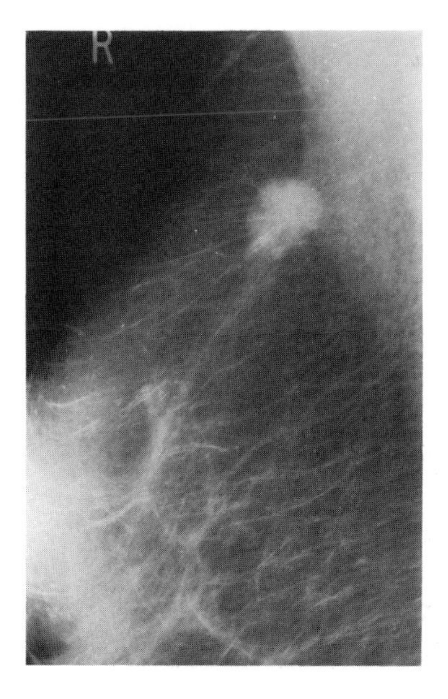

Mammogram showing breast cancer.

Early studies comparing women screened by mammography with control women not so screened found that breast cancers detected in the screened group tended to be smaller, and less often involved the lymph nodes, than those which came to light by self-detection. Ultimately, 30 per cent fewer lives were lost to breast cancer among the screened group. However, the benefit of screening appeared to be confined to women over 50, and even with yearly mammography some women in the screened group developed aggressive cancers in between the screening intervals, somewhat reducing the overall benefit but by no means abolishing it.

More up-to-date studies have confirmed these results, and have even suggested some benefit of mammography in women aged 40 to 50. However, this may have been because of detection of tumours during the clinical examination done with the mammography rather than by the mammogram itself. In women under the age of 50, more than half of all breast lumps are detected by clinical examination and only 25 per cent by mammography. One large study recently suggested an increased death rate in women under 50 screened by mammography, but the reasons for this remain unclear.

In the UK a marked reduction in the number of women with aggressive or lymph node positive tumours has been observed among screened women, and between 30 per cent and 50 per cent of tumours now measure less than 2 cm when they are first diagnosed, whereas 20 years ago this proportion was only 5 per cent to 10 per cent. Even with cancers detected by mammography, the smaller the tumour the better the outcome in general. One study has shown that women with tumours of less than 1 cm in diameter detected at mammography had a 95 per cent survival rate at ten years.

There seems no doubt that screening by mammography can save lives, and it has now been shown to reduce the overall death rate (mortality) from breast cancer by between 20 per cent and 30 per cent among women who have ever been screened compared with others of the same age in the community. It might seem inconceivable that its use could be anything but a good thing. However, there are some serious reservations about a widespread screening programme.

Problems of screening by mammography

Women attending for screening may not be strictly representative of the female population as a whole, which could bias the results of trials. Only about 70 per cent of women invited to attend take up the offer of mammography. Those who do not may be too anxious about what might be discovered, or simply cannot be bothered with the inconvenience involved.

There is some evidence that those who do attend have more breast symptoms than expected, suggesting that some women may be using screening as an opportunity to obtain an opinion on something which had already been bothering them.

Being screened and having a negative result could create a false sense of reassurance and discourage women from examining their breasts in between mammograms. The test itself is not infallible, and a proportion of mammograms may miss cancers which are present (false negative results). Current estimates suggest that 2 or 3 in 10 cancers may not be picked up on mammography.

Alternatively, the mammogram (or the accompanying clinical examination) will pick out as suspicious an abnormality which is not cancer (false positive result, reckoned to occur in about one in 100 mammograms), and this will obviously generate unnecessary worry. If an abnormality is found, it must be further investigated, and inevitably this will involve some biopsies (perhaps half of those generated from mammography programmes) being performed which might not otherwise have been necessary.

A further nagging doubt concerns the increasing number of abnormalities of marginal or uncertain significance detected by mammography as screening programmes have become more widespread. Firstly, the laboratory diagnosis of cancer is not always cut and dried and doctors do not always agree on the significance of borderline changes or on how they should be treated. Furthermore, the natural history of breast cancer is such that some of the cases picked up on mammography would, if left alone, have progressed only very slowly if at all.

Secondly, the detection of *in situ* carcinomas on mammography poses a real dilemma. These represent about 20 per cent of all screen-detected cancers, and the number of new cases of ductal carcinoma *in situ* detected has risen between three- and four-fold in populations served by screening centres. Since few *in situ* carcinomas lead to clinical symptoms, they were rarely diagnosed before the introduction of mammography, and their natural history is largely unknown.

The best form of treatment for these pre-invasive forms of cancer remains uncertain, so if doctors operate as if these women had invasive breast cancer, perhaps even performing mastectomy, there is a risk that many women will undergo unnecessarily extensive surgery. On the other hand if they opt for lesser procedures there is a risk that some cases of *in situ* cancer will become invasive and that these women's chance of cure will have been reduced by not performing definitive surgery earlier.

The potential hazard for large numbers of otherwise healthy women is perhaps the most serious criticism of any screening programme. For every case of breast cancer detected, perhaps hundreds of women must go to the time and trouble of the initial examination, and one or two will face the considerable psychological stress of a biopsy and perhaps surgical procedures for benign disease. While these are, in medical terms, minor procedures, inevitably when large numbers are involved a small percentage will fall foul of surgical or anaesthetic mishaps. In addition, in any large-scale screening programme a number of breast operations, perhaps including mastectomies, may be carried out for tumours which might never have troubled the patient had they been left undetected.

Another consideration is the effects of X-rays on the breast. The doses of X-rays involved in mammography are minute. It has been calculated that the risk of developing breast cancer after one mammogram is less than one in a million. This makes the risk of death from the procedure equivalent to travelling for 30 miles in a car, smoking half a cigarette, or being a 60-year-old man for 10 minutes.

Nevertheless, there is no doubt that radiation can cause breast cancer, that there is a slight risk even with very low doses and that the risk increases with increasing exposure. There is little point in a single screening, and once a mammography programme has been established, women should be invited back at regular intervals for the rest of their lives. Cumulative doses of X-rays over perhaps 20 or 30 years in a large population of women could produce a very small increase in the number of cases of breast cancer. While increased deaths due to this would be infinitesimal in comparison with the number of lives saved by the procedure, it is nevertheless a negative factor to put into the equation.

The optimum scale of mammography screening is also in dispute. Women over the age of 50 may benefit from screening every two years (the current UK programme recommends 3-yearly screening) whereas most studies have failed to demonstrate much benefit in younger women.

Thus the experts still disagree about which women should be screened and how often, about the best means of screening and the treatment for small tumours detected on mammography.

Finally, any screening programme to a degree turns an otherwise well population into 'patients' undergoing medical tests, aware of the possibility of ill-health and being labelled as diseased. The long-term consequences of this are unknown and possibly unmeasurable.

Cost-effectiveness of mammographic screening

All these procedures of course incur time and expense, for patients and their employers, and for the health service and its staff. In a world of finite resources, the total cost per life saved by mammography must be balanced against the alternative benefits which might be available if the money and resources involved were spent elsewhere.

The cost per life saved is not the same as the cost per cancer detected, since not all tumours picked up on mammography will be curable. Furthermore it is necessary to estimate the number of curable cancers detected by screening in excess of those which might have been self-detected, and to correct the figures for the extra cost of treating more advanced cancers which otherwise occur in an unscreened population. These calculations are therefore quite complex.

The cost per life saved is undoubtedly enormous, and has been estimated at £30–40 000. However, aside from arguments about whether it is possible to put a value on, say, the life of a young woman bringing up children, this does turn out to be rather better 'value for money' than, say, screening for cervical cancer.

Also, no one has ever bothered to calculate the cost-effectiveness of saving people from dying in fires or drowning at sea, and quite probably if the same calculations were applied to the Fire Service or the Royal National Lifeboat Institution the cost per life would come out considerably greater than that of breast cancer screening.

The UK screening programme

In the UK a Department of Health working party under Sir Patrick Forrest carefully examined arguments for and against a programme of mass screening and recommended, in the 1987 Forrest report, that a national breast cancer screening programme be set up, funded by the NHS.

This offers all women aged 50 to 64 screening by mammography at three-yearly intervals, with the same examinations available to older women on request. The aim is to reduce breast cancer deaths in the screened population (who represent just over half of all breast cancer deaths) by 25 per cent by the end of the century. This would represent some 2000 lives saved per year.

Results of the first round of screening show that 71.4 per cent of women invited for screening actually attended. Of these women, 7.1 per cent were recalled for further investigation, 1.0 per cent underwent a biopsy, and 0.63 per cent were found to have cancer. Among the cancers detected, 17.6 per cent were non-invasive *in situ* duct carcinomas, and among the invasive cancers 20.9 per cent measured 10 mm or less in diameter.

Mammography in diagnosis

As well as being used as a screening procedure to detect cancer in symptomless women, mammography can also be useful in some circumstances as an aid to diagnosis of breast symptoms, particularly where there is no obvious single lump (see Chapter 4). It is also used in the follow-up of women who have had successful treatment for breast cancer.

Use of mammography for localization

If a mammogram reveals a pattern suspicious of cancer but not an actual cancerous lump, a needle localization biopsy may be needed to find and biopsy it, enabling surgery to proceed with the removal of as small a quantity of breast tissue as possible. Accurate localization makes the operation simple and fast, decreases the complications, and yields a better cosmetic result.

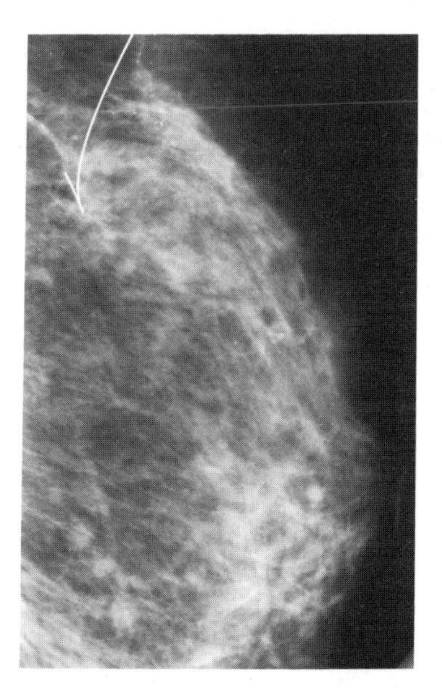

Needle localization biopsy aided by mammography.

First, a radiologist (a doctor specializing in X-ray techniques) gives a small injection of local anaesthetic and inserts a barbed needle into the quadrant of the breast containing the suspicious area, as determined by two mammographic views taken at right angles to each other from the top and the side of the breast. Further X-rays are taken if necessary as the tip of the needle is adjusted to lie close to the suspicious area. The needle is then advanced so that its barb escapes from the sheath and is hooked securely in place.

Meanwhile the surgeon is standing by in a nearby operating theatre, and once the needle is in place the woman is given a general anaesthetic. The surgeon then cuts down on the needle and cuts out a small amount of tissue from around its tip.

This specimen, with the needle, are sent back to the X-ray department where more pictures are taken to confirm that the suspicious area has indeed been removed. If so, the surgeon can go ahead and sew up the wound.

If not, additional samples of tissue can be taken and the procedure repeated. The sample is then sliced and stained to be examined at leisure in the laboratory. A woman may have to wait 48 hours for a firm diagnosis under these circumstances, which is obviously a period of intense anxiety but does avoid unnecessarily extensive surgery.

Occasionally further mammograms will be needed once the woman wakes up from the biopsy procedure, to make sure that no suspicious areas have been left behind.

Breast self-examination

Since over 90 per cent of breast tumours are detected by the woman herself, it might seem obvious that regular and systematic examination of her breasts would increase a woman's chance of finding a breast cancer as soon as possible after it becomes detectable.

Certainly, a woman who is aware of what her breasts normally feel like is the best person to detect any changes. Particularly for women with naturally lumpy breasts or those whose breasts change their feel greatly during the menstrual cycle, self-examination may be a better means to detect early changes than examination by a doctor.

There have been several major programmes to teach women breast self-examination, and urge them to check their breasts once a month, at the same time in their cycle. However, despite much enthusiasm for self-examination, and some evidence that it does lead to earlier detection of disease which makes treatment easier, there is no proof of any impact on breast cancer death rates.

Why this should be so is curious, but as with screening there are several reservations about breast self-examination. Firstly, one survey has shown that 70 per cent of women with breast lumps intentionally delay reporting their symptoms to their family doctor. This delay varied from a few months to several years. As discussed in Chapter 10, women seem to delay not through ignorance of the significance of breast lumps, but through fear of a possible diagnosis of cancer. Nearly half of all the breast cancers seen for the first time in hospital clinics are already at an incurable stage, with many tumours more than 5 cm in diameter, which must have been easily detectable for some time.

Clearly breast self-examination will achieve nothing if women do not act on what they find. Promoting a greater awareness that most breast lumps are not cancerous might encourage more women to visit their doctors early on.

Another possibility which has been suggested is that monthly examination becomes a ritual which gives women a false sense of reassurance, may discourage them from conventional screening, and perhaps generates anxiety or obsessive behaviour (examining the breasts daily instead of monthly, for example). In addition, widespread promotion of self-examination could make women who do not practise it regularly feel guilty if they do develop a tumour.

It may also be that the difference in stage between tumours detected by self-examination and those found in the normal course of events is so small that it makes little impact on the chance of surviving a breast cancer.

A large randomized trial in Leningrad in the former USSR comparing tumours which arise among women taught breast self-examination with those among a comparable group of women not given this teaching has just been completed. This showed that although tumours were detected earlier in the self-examination group, there was no difference in breast cancer survival rates between women who practised self-examination and those who did not, and the group practising self-examination had twice as many biopsies for benign disease.

Meanwhile in the UK the focus of the campaign has recently changed from one of regular self-examination to one of 'breast awareness'. Again, this is based on the premise that the woman is the best person to detect any changes in her own breasts, and is intended to encourage women to become familiar with their breasts and to look for change in a non-ritualistic way, learning to distinguish between normal fluctuations and abnormal change, and seek medical advice for the latter as soon as possible. The best and simplest way to do this is to feel the breasts with a soapy hand (rather than a flannel) while washing. Whilst it cannot at this stage be guaranteed that this will produce any gain in life expectancy if a breast cancer is detected, it will allow prompt reassurance in most cases and, where a tumour is present, increases the chance that conservative surgery will be possible.

Breast changes for which medical advice should be sought

- dimpling or flaking of the skin
- nipple discharge
- lump or thickening in the breast tissue
- unusual pain or discomfort
- any NEW difference between the appearance of the breasts (when looking, lifting them, or moving the arms)

14

Concluding thoughts

Anyone asked to predict a major advance in the treatment of breast cancer for the next millenium might reasonably respond by suggesting a new type of drug based on a better understanding of the process of cancer development.

There are certainly several lines of research now being undertaken in the field of cancer (oncology) generally which are pursuing such a goal with vigour. One promising avenue is to try to swing the balance between cancer cell invasion and the body's defence mechanisms more in favour of the patient.

This could in theory be attempted by boosting the direct attack on cancer cells mounted by certain substances produced naturally in the body in response to infection or malignancy. Interferon is one such agent, and when it was first made available in sufficient quantities to be tested in treatment generated great enthusiasm, but hopes that it might be a general anti-cancer agent have now abated. Although it does undoubtedly have benefit in certain malignant conditions it is by no means the cure-all anticipated and, as far as breast cancer is concerned, there is no good evidence that it has any effect which could not be achieved with much less expense using conventional chemotherapy.

Another variant on this line is to exploit features of the body's own immune response to try to produce immunity to cancer, in a similar manner to immunization against infectious diseases. This is known as immunotherapy. Examples are the use of the fearsome-sounding 'lymphokine activated killer (LAK) cells', and heat shock proteins, recently produced by genetic engineering, which may act to stop tumours developing.

However such approaches, although undoubtedly sophisticated, may suggest a degree of tunnel vision. They are based on the premise that the extent of spread of breast cancer cells (micrometastases) at the time of diagnosis is the crucial determinant of outcome of treatment, and that this spread arises from cells shed from the primary tumour and travelling via the bloodstream early in the course of the disease. While this theory is attractive, and has generated new therapeutic approaches which promise improved survival, it may be only part of the truth.

For example, an alternative hypothesis to explain the phenomena observed in the course of breast cancer treatment might be that secondary tumours do not arise as a result of spread of intact cancer cells capable of reproducing themselves in distant organs. Rather, the cancer cells could shed genetic material, DNA, which has mutated and is effectively foreign to the body, and which then selectively 'infects' other susceptible cells. For example, such DNA might alter the genetic make-up of a normal liver cell to force it to behave like a breast cancer cell. Such a hypothesis could go some way towards explaining why secondary tumours are common in some tissues such as liver and bone marrow, but almost never occur in others, such as muscles.

So an alternative approach might be to try to 're-educate' cancer cells to behave like normal cells, instead of crude attempts to kill them off. This might be achieved through some kind of gene therapy, or through alterations to the various growth factors which influence cell behaviour.

These examples indicate the need for scientists and clinicians to keep an open mind to novel ideas, even if they challenge previously cherished beliefs. However, an understanding of the fundamental nature of breast cancer, while vitally important to future progress, is far from the sole answer.

Experience suggests that a breakthrough in cancer treatment in the nature of a 'magic bullet', like antibiotic treatment for bacterial infections, is unlikely to be achieved in the near future however much wishful thinking is poured in this direction. Cancer is too complex a process and progress is likely to depend on broad-based advances on many fronts.

As we have seen, improvements in breast cancer survival have often come about as much from better selection of patients for different treatments as from improvements in the treatments themselves. The discovery of oestrogen receptors in certain tumours has not proved as helpful a predictor as was first hoped, but it is possible that other biological tumour markers now uncovered, such as growth factor receptors or genetic markers, or new factors awaiting discovery, will reveal more about the nature of an individual tumour and allow better selection of optimum therapy. Monoclonal antibody technology may enable doctors to assess tumour load and disease stability and so to offer a better prediction of survival for individual women.

However, scientific advances by themselves are not enough. Laboratory research findings and the hypotheses they generate must in the end translate into improved survival for women with breast cancer, and the proof of this can only be achieved by painstaking analysis of data from large clinical trials.

The difficulties posed by the need for clinical trials have been touched on elsewhere in this book, and represent one of the most formidable problems which cancer researchers must tackle. Formal trials are the only way to be certain that new initiatives represent real advances and that future patients can be offered treatment in confidence that it is the best available. We have no better means of testing new treatments, assessing short- and long-term outcomes, or guarding against unanticipated effects of therapy.

Yet in a sense the randomized controlled trial which is the scientific ideal is ethically impossible. It requires that any doctor asking a patient to participate sincerely views the two alternatives being tested as having equal potential therapeutic value for that patient. It requires that the patient, having been provided with all the facts, and understanding them, also has no preference whatsoever for one form of treatment over another. The likelihood of both doctor and patient feeling this way is remote.

In some ways therefore clinical trials require patients to make a selfless gesture on behalf of future sufferers from their particular disease. It might be said that patients have a moral obligation to participate, since otherwise they gain from scientific advances only at the expense of others undertaking the risk on their behalf.

On the other hand, it is ethically dubious to try to coerce people into 'volunteering' for trials. It is an ethical imperative that people should only enter a trial if they understand all the factors involved. Yet to bombard someone who has just been diagnosed as having a life-threatening illness with unsolicited information about complex issues, which even doctors may not fully understand, might be regarded as cruel. To seek fully informed consent for such a trial could in some circumstances conflict with a physician's duty to hold paramount the best interests of the patient.

It is vital that we face up to this extraordinary dilemma. One possible solution could be to establish a network of informed, socially responsible women committed to doing their bit for improving the outcome for women's cancers. They could form a special interest group to give impetus to initiatives to fast track the evaluation of promising new treatments for cancer.

These women would be educated about 'female' cancers (such as those of breast and cervix) and about the need for controlled trials, and could spread the message to others. They could be kept up-to-date with current treatments and listings of trials and their scientific rationale. They would recognize the personal benefits of being involved in a trial. Participants generally fare better, irrespective of which trial group they are allocated to, than patients with the same disease not in a trial, probably because they get more attention. They are more likely to be cared for by experienced specialists committed to their particular disease, to be treated in specialist centres, and to receive more regular monitoring.

If large numbers of currently well women were recruited in this way, inevitably in time some would develop the diseases in which they had an interest. They would not then, at the time of their diagnosis, have to absorb masses of information about their condition, the treatments on offer, and the need for randomization, because they would already be well-informed.

They would be well-motivated to take part in trials and might indeed expect to be offered the chance to do so.

One means of establishing such a network, perhaps Europe-wide, would be to divert the energies of various national and voluntary bodies into promoting the benefits of randomized controlled trials. Such an initiative would serve women better than many current attempts to promote breast self-examination, now known to be a futile procedure, or to expand screening programmes to inappropriate age groups.

Given that breast cancer is often a slowly progressive tumour, improvements in the quality as well as the quantity of life for affected women represent major progress. The current trend towards increasingly conservative surgery is likely to continue and to stimulate demand for more facilities for reconstructive surgery, so that fewer women will have to have mastectomy and those who do will not have to face life without a breast as well as with their cancer.

'Kinder' surgical techniques and improved understanding of the psychological sequelae of cancer and of breast cancer treatment could have another spin-off. If orthodox medicine presents a more humane face, fewer women may be tempted by the lures of unscrupulous practitioners of fringe medicine. It is understandable that panic and despair can lead some women to desperate measures, but transglobal searches for miracle cures merely risk wasting what time remains on what often becomes an increasingly costly obsession.

On the other hand, co-operation between medicine and reputable complementary medical practitioners increasingly enables techniques which are useful to be made available to patients without denying them the benefits of orthodox treatment. An example would be the use of acupuncture to alleviate nausea during cancer chemotherapy. Such co-operation can allow the best of both worlds while ensuring that the public is not waylaid into futile or expensive pie-in-the-sky therapies.

Perhaps the best immediate hope of improving the outcome of breast cancer is to increase the proportion of tumours detected at an early stage.

The current breast awareness campaign and continued public

education and enlightened media discussion may make inroads into the psychological blocks which cause women to delay presenting their breast lumps to a doctor.

Analysis of the results of the UK screening programme will be a critical test of the applicability of mammography. The aim of reducing breast cancer deaths among the age group invited for screening by a quarter by the year 2000 is laudable, but equally important will be the assessment of screen-detected non-invasive tumours and decisions about the management of abnormalities whose significance is uncertain.

In addition, research aimed at better means of predicting and testing for breast cancer risk needs to be accompanied by a consensus on the best way to monitor women at high risk, and what prophylactic strategies—such as the use of tamoxifen or preventive surgery—are justified in otherwise healthy women.

Finally, the ideal goal would be to prevent breast cancer development in the population at large. If a drug could be discovered which was both a safe and effective contraceptive and protected against breast cancer, this would have enormous benefits for womankind. Such an idea is not far-fetched, since a 'Pill' type hormone which worked by reducing the production of oestrogens from the ovaries could provide this double benefit.

One possibility now being explored is a group of drugs which mimic the actions of hormones produced in the brain to regulate the synthesis of hormones in the ovaries.

These could be given with small amounts of oestrogen and progesterone to reduce the menopausal side-effects which might otherwise occur, and offer added protection against endometrial and ovarian cancer, as is provided by the contraceptive Pills in use today.

Another is the use of a drug called gestodene which is currently given as the progestogen component of some newer types of combined contraceptive Pills. Preliminary evidence suggests that it may help to prevent breast cancer.

In conclusion, it is clear from many of the avenues discussed that a blind search for scientific breakthrough is not by itself sufficient. If we are to avoid some of the mistakes of the past century which have stemmed from rigid adherence to dogma and obsolete traditions, we need to ensure that new hypotheses do not generate new dogma to inhibit progress. This is particularly so if decisions about such areas as research funding or national policies on screening span the political arena. The capacity for rational decision-making in the light of current medical understanding is at least as important as the increased knowledge gained through research itself.

Further reading

John Hinton (1972). *Dying*. Penguin, London.

Elizabeth Kubler-Ross (1970). *On death and dying*. Tavistock, London.

Richard Lamerton (1980). *Care of the dying*. Pelican, London.

Robert Buckman (1988). *I don't know what to say. How to help and support someone who is dying*. Macmillan, London.

When someone with cancer is dying. Pamphlet available from Cancerlink (see Appendix for address).

Lesley Fallowfield and Andrew Clark (1991). *Breast cancer*. The experience of illness series. Tavistock/Routledge, London.

Karin Gyllensköld (1982). *Breast cancer, the psychological effects of the disease and its treatment*. Tavistock, London.

June Marchant (1978). *Rehabilitation of mastectomy patients*. Heinemann, London.

Appendix

Hospice Information Service
St. Christopher's Hospice
51-59 Lawrie Park Road
London SE26 6DZ
Tel. 081 778 1240
Directory of hospices and home care support teams available for cancer patients throughout the UK.

Women's National Cancer Control Campaign
Suna House
128-130 Curtain Road
London EC2A 3AR
Tel. 071 729 2229 (9:30am-4:30 pm weekdays)
Recorded breast screening advice Tel. 071 729 4915 (24 hrs)

Breast Care Campaign Helpline 0628 481233 (5pm-8pm)

Cancerlink
17 Britannia Street
London Wc1
Tel. 071 833 2451

BACUP
3 Bath Place
London EC 2
Tel. 071 613 2121

Breast Care and Mastectomy Association
Anchor House
Britten Street
London SW2 3T2
Tel. 0500 245345

Index

Bold numbers denote reference to illustrations.